MICROCOSMS

Sacred Plants of the Americas

JILL PFLUGHEBER
STEVEN F. WHITE

MICROCOSMS

Sacred

Plants

of the

Americas

In the memory of my late sister, Gina Wells (9th February 1955 – 3rd February 2022), a Biology major at St. Lawrence University with a special passion for botany.
~ Jill Pflugheber

For Esthela Calderón, whose words have joined mine "so dreams can wander and flower in one of the world's desolate places".
~ Steven F. White

First published in Great Britain in 2025

An imprint of Academy Editions Ltd
Kimber Studio, Winterbourne, Berkshire, RG20 8AN, UK
info@papadakis.net | www.papadakis.net

  @papadakisbooks

Publisher: Alexandra Papadakis
Design: Alexandra Papadakis
Production: Alexandra Papadakis & Molly Dewar
Text editing: Molly Dewar
Map illustrations: Molly Dewar
Publishing assistant: Jemima Lane Fox

ISBN 978 1 906506 79 7

Copyright © 2025 Jill Pflugheber, Steven F. White, and Papadakis Publisher.
All rights reserved.

Jill Pflugheber and Steven F. White hereby assert their moral right to be identified as authors of this work.

All text and images © Jill Pflugheber and Steven F. White unless otherwise stated.

No part of this publication may be reproduced or transmitted in any form or by any means, electronic or mechanical, including photocopy, recording, or any other information storage and retrieval system, without prior permission in writing from the Publisher. No part of this publication may be used or reproduced in any manner for the purpose of training artificial intelligence technologies or systems.

A CIP catalogue of this book is available from the British Library.

Printed and bound in China.

Papadakis is committed to a sustainable future for our business, our readers, and our planet. This book is made from Forest Stewardship Council™ certified paper, responsibly sourced from forests that are managed in an environmentally and socially responsible manner.

front cover: *Ceiba pentandra* (ceiba); confocal image
back cover: *Brugmansia aurea* (golden angel's trumpet); plant
endpapers: Kené artwork by an anonymous Shipiba artist
frontis: *Banisteriopsis muricata* (red ayahuasca); confocal image

The *Microcosms* website and digital herbarium is accessible in five languages.
For more information please visit https://www.microcosmssacredplants.org/

The End of the Corn Festival,
by contemporary Wixárika artist Antonio López Pinedo

"When the Corn Festival is over, the person who hosted it sends some of his family members (or goes himself) to make offerings to the gods, which they carry in knapsacks and gourds for sacred water. And they leave on their journey to the gods, like the god of the sea (the snake), to the caves in the mountains, to the temples like tiny homes, to Real Catorce (Wirikuta), the sacred place of the peyote, the rabbit sent by the sun, and, above all, the antlers of the sacrificed deer."

CONTENTS

Introduction	9	*Steven F. White*
Microcosms: The Inner Worlds of Sacred Plants	21	*Steven F. White*
Alicia anisopetala	35	Black ayahuasca
Amaranthus cruentus	39	Amaranth
Anadenanthera colubrina	43	Cebil
Artemisia ludoviciana subsp. *mexicana*	51	Western mugwort
Ayahuasca / Yagé:	55	
Banisteriopsis spp.	55	Ayahuasca / yagé
Diplopterys spp.	55	Chaliponga / huambisa
Psychotria spp.	55	Amyruca / chacruna
Bourreria huanita	79	Esquisúchil
Brugmansia insignis	83	Angel's trumpet
Brugmansia × *candida*	83	Culebra borrachero
Brugmansia sanguinea	83	Blood-red angel's trumpet
Brunfelsia grandiflora	95	Chiric-sanango
Bursera fagaroides	101	Copal
Calea ternifolia	105	Zacatechichi
Cannabis sativa	109	Marijuana
Ceiba pentandra	115	Ceiba
Cestrum parqui	121	Palqui
Datura innoxia	125	Toloache
Desfontainia spinosa	129	Taique
Dianthera pectoralis	135	Carpenter bush
Drimys andina	139	Dwarf winter's bark
Erythroxylum novogranatense	143	Coca

Heimia salicifolia	147	Sun-opener
Hierochloe odorata	151	Sweetgrass
Ipomoea corymbosa	155	Ololiuhqui
Ipomoea tricolor	155	Heavenly blue morning glory
Latua pubiflora	161	Latué
Leonotis nepetifolia	163	Christmas candlestick
Lophophora williamsii	167	Peyote
Mimosa pudica	173	Sensitive plant
Mimosa tenuiflora	177	Tepezcohuite
Neltuma spp.	181	Mesquite
Nicotiana rustica	189	Tobacco
Paullinia cupana	193	Guaraná
Polylepis incarum	197	Queñua
Psilocybe cubensis	203	Di-shi-tjo-le-rra-ja / magic mushroom
Salvia apiana	207	White sage
Salvia divinorum	209	Diviner's sage
Solandra maxima	215	Chalice vine
Tabernaemontana sananho	221	Uchu sanango
Tabernaemontana undulata	221	Mana heins
Tagetes lucida	225	Mexican marigold
Theobroma cacao	229	Cacao
Trichocereus macrogonus var. *pachanoi*	235	Huachuma / San Pedro cactus
Ullucus tuberosus (*aborigineus*)	239	Olluko / ancestral potato
Virola theiodora	245	Yãkoana
Zea luxurians	251	Guatemalan teosinte / wild maize

About the Authors	255
Index of Plants Illustrated	256
Bibliography	258
Acknowledgements	262
Picture Credits	264

Native distribution

Introduced to

Introduction

Steven F. White

Confocal microscopy, also known as confocal laser scanning microscopy, is a specialised optical imaging technique that provides contact-free, non-destructive measurements of three-dimensional objects.

For this book, plants considered sacred by Indigenous groups of the Americas were scanned at St. Lawrence University's Microscopy and Imaging Center.

The procedure gathers information from a narrow depth of field, while simultaneously eliminating out-of-focus glare and creating optical sections *through* layers of biological samples. Images are built over time by gathering photons emitted from fluorescent chemical compounds naturally contained within the plants themselves, creating a vivid and precise colorimetric display.

To pay homage to sacred plants revered by Indigenous groups throughout the Americas is a way of honouring the entire world in a time of environmental emergency. This book – at the juncture of art, technology, and science – magnifies life in ways that may alter how humans perceive other living entities from our shared and threatened biosphere in more egalitarian terms.

The plants reveal themselves as 21st century extensions of biomorphic forms that were the genesis of abstract works by artists such as Wassily Kandinsky and Paul Klee one hundred years ago.

Some of these plants contain the most potent psychoactive agents on the planet, and serve as intermediaries that have enabled Native communities to communicate with their ancestors, wage war on the enemies of their land and their traditions, conceptualise entire cosmogonies, and maintain a nearly impossible ecological equilibrium.

Each stoma, each trichome, each patterned fragment of xylem and vascular tissue as well as each grain of pollen in these vital portraits is not only a way into previously unseen vegetal realms, but also a potential way out of our collective crisis.

VISIONARY ART AT ST. LAWRENCE UNIVERSITY
This book reproduces and significantly expands the exhibition *Microcosms: A Homage to Sacred Plants of the Americas* that opened at St. Lawrence University's Brush Art Gallery on 2nd March 2020, and, like so much else around the world at that time, was forced by the Coronavirus pandemic to close prematurely only two weeks later. Even so, the deadline for the preparation of this event, some four years in the making, obliged Jill Pflugheber and I to consider the space limitations of the gallery and then select with artistic and scientific rigour approximately 50 images of some 35 different plant species for the show, printing them in a large format (457 × 457mm / 18" × 18").

György Kepes, Exhibition panel no. 17 (Cross section of root of redwood: 100×)

The *Microcosms* book is a natural extension of two prior exhibitions at St. Lawrence University (in addition to the 2020 *Homage to Sacred Plants of the Americas*): *Visions that the Plants Gave Us* and *Inner Visions: Sacred Plants, Art, and Spirituality*, both curated by Professor Luis Eduardo Luna, who is the director of Wasiwaska, a research centre for the study of psychointegrator plants, visionary art, and consciousness in Florianópolis, Brazil. He was also named Doctor of Humane Letters by St. Lawrence University in 2002.

These previous exhibitions gathered visual art by numerous national and international artists, including work by Indigenous creators who identify themselves as Cashinahua, Huichol (Wixárika), Huni Kuin, Shipibo, Siona, and Witoto. *Inner Visions* opened with an extensive showing of Donna Torres's precise and elegant botanical drawings of many of the same plants that appear here in the *Microcosms* book. A series of confocal images are available via the *Microcosms* digital image collection on JSTOR.

PLANTS OF THE GODS

We understand "sacred" in an ample way, in the reverential and respectful sense that Amerindian groups define this term as a spiritual pact, and have included a wide (albeit still limited) range of plants from Maize to Peyote, from Amaranth to the plants used to prepare ayahuasca, from the Mapuche's Foye to the Yanomami's Yãkoana, and from an Incan ancestral potato Olluco to the San Pedro cactus. There is also a bonus image of the obligatory fungus known to Indigenous Mesoamerica as teonanácatl (flesh of the gods) to accompany all these plants. The texts that describe these species clarify Amerindian medicinal and spiritual uses associated with them. More often than not, the revered plants that appear in this collection are psychoactive. Why? According to the great Harvard ethnobotanist Richard Evans Schultes and his co-author Albert Hofmann, the Swiss scientist who was the first to synthesise LSD: "plants that alter the normal functions of the mind and body have always been considered by peoples of non-industrial societies as sacred, and the hallucinogens have been *plants of the gods* par excellence […] It is in the New World that the number and cultural significance of hallucinogenic plants are overwhelming, dominating every phase of life among the aboriginal peoples".

DIGITAL ART AND TOOLS FOR SEEING

These images of plants held sacred by Indigenous groups of the Americas can also be viewed within the critical framework of *Microcosmic Phytoformalism*, a term that actually did come to me in a dream, as cliché as this may sound, at 3:00am on 12th February 2020. I scrawled out the letters in two quavering lines in a notebook with a pen in the darkness and went back to sleep. In my grateful mind, it was thrillingly perfect, and I hope that the gathering of these confocal images allows this newly-coined designation to germinate like a seed into an organic structure that will help us understand *what* and *how* we see. From its technological origin, it represents a new stage of an ever-evolving history of both microscopy and art, revealing colours, shapes, and textures combined in a compelling, growth-related vision derived from living biological materials.

Here in this book, then, selected from (literally) terabytes of what my science-oriented colleague cannot help but refer to as "data", is a harvest of confocal plants for visualising a natural order that has existed all along, even if it has remained less than perceptible until quite recently.

According to Hungarian Bauhaus artist and MIT professor György Kepes, whose pioneering work which explores the connections between art and science is an important precedent, "a pattern in nature is a temporary boundary that both separates and connects the past and the future of the processes that trace it". Every pattern, he says, is a "space-time boundary of energies in organisation".

Like the Hubble telescope that has produced so many iconic celestial images, the confocal microscope is a tool of perception that extends humanity's narrow biological filters. And now, we've used our human eyes and technological vision to design the James Webb Space Telescope so that we are able to perceive the universe in colours that no human eye has ever seen. Perhaps this kind of *seeing beyond* parallels the neurological effects of these plants of power themselves as they have crossed blood-brain barriers and exercised their profound influences in ritual contexts, in some cases, for millennia.

Is it possible to imagine these images as art that breathes? Could we breathe with the stomata that appear before our eyes? Does this microcosmic art reflect biological processes that allow human beings to participate with plants in a *co-becoming*? Is this an example of how the infinitely small begins to approximate the infinitely vast, the way recent images of the sun show a surface that resembles myriad kernels of corn, each of which is the size of the state of Texas?

In these inspiring microscapes, born of an art-science symbiosis, there is sometimes an intentional preference (especially on my part) not for perfect, unscathed, whole forms, but rather the beauty in a broken trichome, a collapsed grain of pollen, ripped vascular tissue, and structures ruptured perhaps by a plant's long clandestine journey across borders and within restrictive systems.

This is transgressive art, an art of resistance. The art, finally, of surviving in a threatening world in which laws and repressive security forces with unchecked power continue to discriminate against plants, harassing, arresting, and imprisoning the people who use them for spiritual and academic purposes. Tragically, the violent fear underlying and fuelling the Spanish Inquisition of time past is still a grave threat in the 21st century around the world! As many have said, the War on Drugs is a War on Consciousness.

Reading Michael Marder's *Plant-Thinking: A Philosophy of Vegetal Life*, I began to ask myself some questions: How can we give a new prominence to the vitality of plants? How is it possible for us to encounter plants and *not* take them for granted? Plants are so absolutely familiar, yet, at the same time, so utterly foreign. We regard plants with what Marder calls an "instrumental attitude", always wondering how we can put them to good material use. But if we were able to move beyond the impediments that we have erected between ourselves as humans and plants, could we somehow turn our utilitarian approach to vegetal lives (in their astonishing variety) into a way of perceiving them differently, "recreating the plant in imagination"?

In this sense, *The Farther Shore: A Natural History of Perception* by Don Gifford, my favourite professor when I was a student at Williams College in the 1970s, provides some fascinating historical, scientific, and aesthetic insights regarding perception as "creative filter", and the importance of having an awareness of *how* we are perceiving, especially in "the mediating presence of optical instruments". A confocal microscope would fit perfectly in the category of tools that help defamiliarise what Gifford calls "an all-too-familiar everyday world". The images of sacred plants as art make the known suddenly, and perhaps shockingly, unfamiliar. As sites of contemplation, they can trigger visionary experiences in ways corresponding to Romantic notions of time that can be found in poets such as Wordsworth. The images, like the poems, inextricably linked to the natural world, are a means of therapeutic seeing: they can become, in Gifford's words, a "fully remembered, and therefore repeatable, sequence of microcosmic visions of eternity". The strange and uplifting confocal art from ephemeral plants and the molecules they contain are portals to what it will take to prevail in the climate war.

Perhaps this book can jolt us into a necessary space of psychological transformation, now, before it is too late. This century's technology, then, can become a means to facilitate paying homage to the incredibly diverse ordering principles of the plant-teachers that are the basis of Amerindian spirituality. Right before our very eyes. Take your time with these images. They will still be here, though perhaps with new meanings, when you return from inner journeys. If you need to know these plants in other ways, they will find you.

In *Art as Organism: Biology and the Evolution of the Digital Image*, Professor of Aesthetic Studies Charissa N.

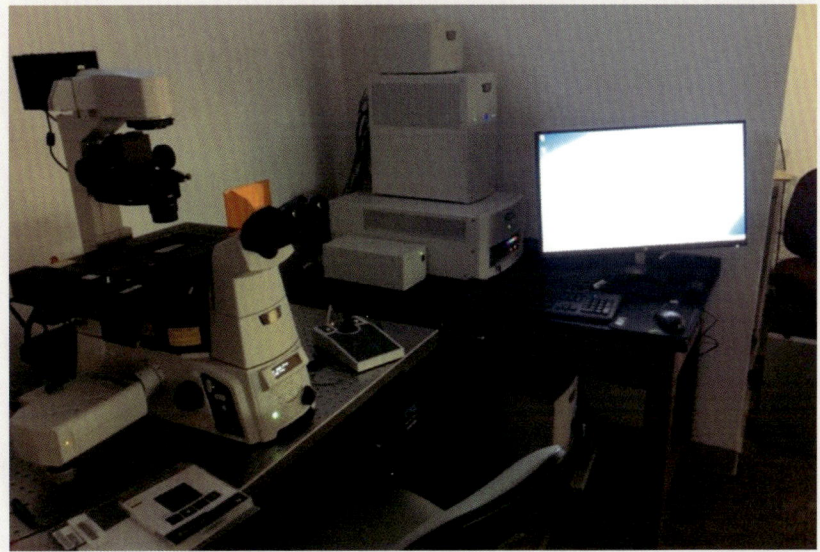

top: Invitation to the exhibition "Microcosms: A Homage to Sacred Plants of the Americas", Richard F. Brush Art Gallery, St. Lawrence University, 2nd March (11th April, 2020)

above: Nikon C2+ confocal microscope

Terranova finds theoretical links between biology and the digital image that coincide perfectly with the goals of *Microcosms – Sacred Plants of the Americas*: "extending outside of art, looping back out into the world in emergent action, this story connects to a bigger politics of ecology, the environment, and radical and rapid climate change – or life in the time of the anthropocene".

ARTISTIC & SCIENTIFIC ANTECEDENTS TO MICROCOSMS

Robert Hooke (1635-1703) is the English author of the landmark book from 1665, *Micrographia: or some Physiological Descriptions of Minute Bodies made by Magnifying Glasses with Observations and Inquiries Thereupon*, the first publication containing descriptions based on observations with the aid of a microscope. He invented the word cell after studying a sliver of bark from a cork tree (*Quercus suber*), and preserved this pioneering knowledge with his own drawing.

Johann Wolfgang von Goethe (1749-1832), the great German poet, invented the word "morphology", which Gordon L. Miller defines as "a science of organic forms and formative forces aimed at discovering underlying unity in the vast diversity of plants and animals". Goethe was also the author of *The Metamorphoses of Plants*, originally published in 1790, a work that transformed 19th century biological thought.

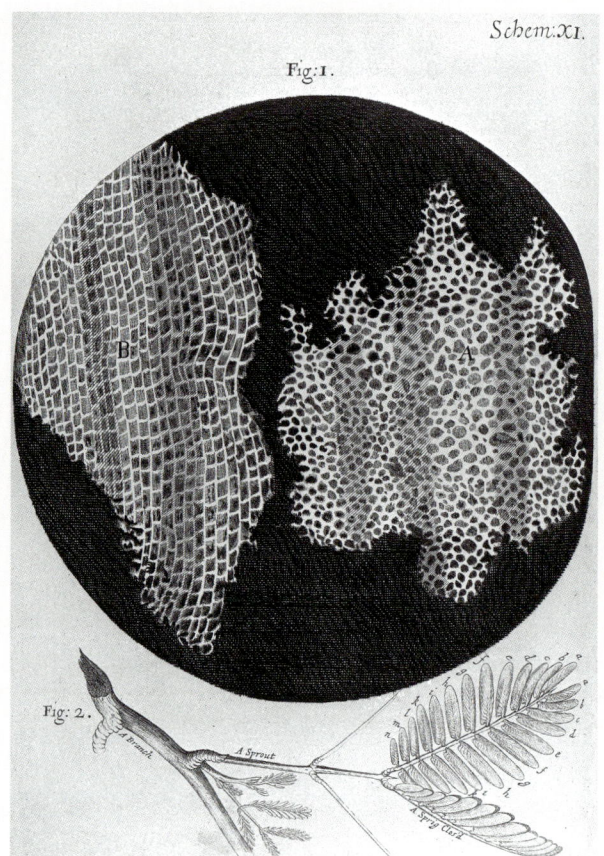

With the improvement in microscopy during his life, Austrian botanical artist **Franz Bauer** (1758-1840) was able to produce exquisitely detailed studies of a wide range of pollen types.

About **Anna Atkins** (1799-1871), the first female photographer and the inventor of the cyanotype, by means of which she created detailed blueprints of algae from Britain that she published in book form in 1843, Larry J. Schaaf maintains that "in the course of a scientific endeavour, [she] turned her 'fondness for botany' into lasting symbols of beauty and expression".

Ernst Haeckel (1834-1919), who coined the word ecology in 1866, is the artist who produced the influential work *Kunstformen der Natur* in 1904. His precise drawings of microscopic radiolarians (1862) were particularly influential in the architectural symmetries that emerged from the Art Nouveau movement, and in *Jugendstil* artists of the late 19th century, thereby linking aesthetics to Darwinian theories of evolution.

In 1923, **R. H. Francé** (1874-1943), who might be characterised as a continuation of the model of the German Romantic scientist, wrote: "it is only in the last thirty years that the microscope has been perfected to the point of spying out the minute and secret structure of the cell". In keeping with contemporary environmental ideas, he believed that "the world is a unity, each part of which influences all the others". Redemption and solutions, he affirmed, could be attained only by acting in harmony with the forces of the natural world. In *Germs of Mind in Plants* (1905), Francé exclaims with unbridled enthusiasm that, after the invention of the achromatic lens and

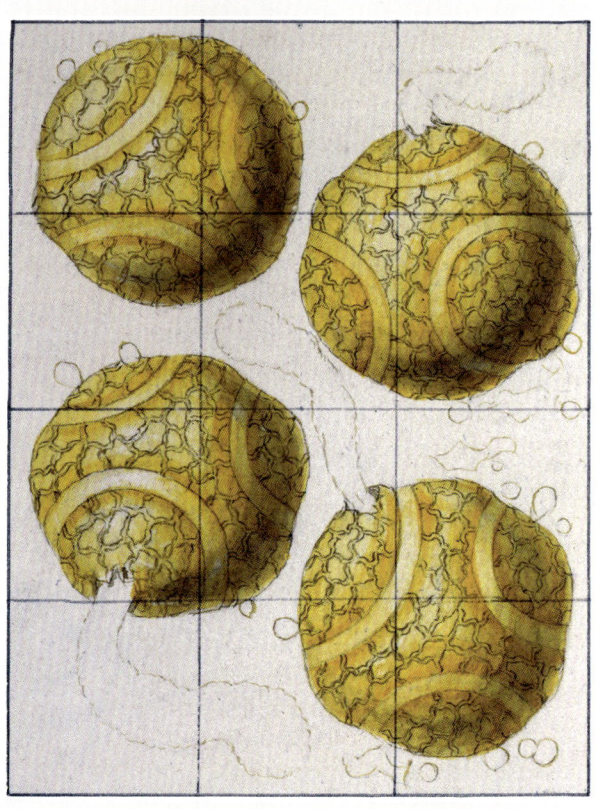

top: Robert Hooke, drawing of *Quercus suber* bark (aided by a microscope) (1665)

above: Franz Bauer, *Passiflora caerulea pollen*, from his collection of illustrations, *Epidermis Floris. Pollen grains, Monstrosities*

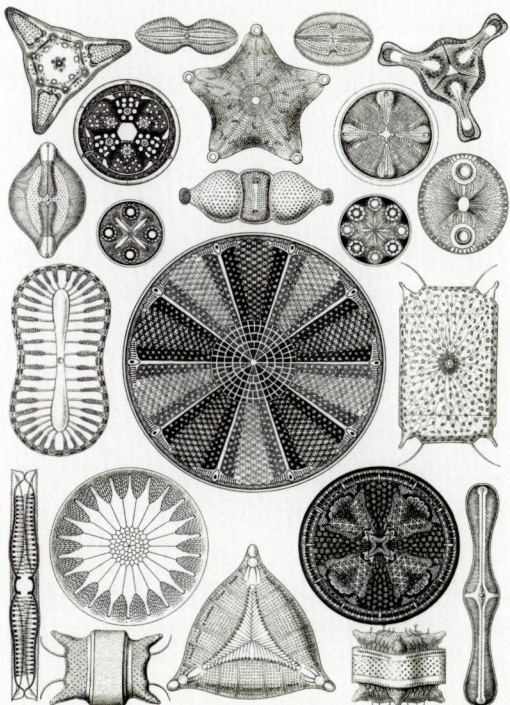

top left: Anna Atkins, cyanotype, *Asplenium braziliense*, South America (1854). Atkins is the first person to publish a book illustrated with photographs (1843)

top right: Diego Rivera, "Man, Controller of the Universe" (detail) (1934)

above: Ernst Haeckel, "Diatomea" from *Kunstformen der Natur* (1899-1904)

microscopy's ability to reveal previously invisible worlds in astonishing detail and colour, "we are now in a certain sense considering the very foundations of knowledge".

Wassily Kandinsky (1866-1944), one of the founders of abstract art whose biomorphic forms are derived from his knowledge of biology, wrote the following in 1935: "this experience of the *hidden soul* in all the things seen either by the unaided eye or through microscopes or binoculars, is what I call the *internal eye*. This eye penetrates the hard shell, the external *form*, goes deep into the object, and lets us feel with all our sense its internal *pulse*". These ideas are abundantly evident in "Colourful Ensemble" (1938), "Striped" (1934), and "Dominant Curve" (1936) from the Solomon R. Guggenheim collection.

Other artists who incorporated biomorphs as an attempt to capture strange new microscopic landscapes in their work include **Hans Arp** (1886-1966) and **Joan Miró** (1893-1983) in his painting "Carnival of Harlequin" (1924-25).

After he became fascinated with the pioneering research on contagious illness published by French microbiologist Louis Pasteur (1822-1895), and began to collaborate with botanist Armand Clavaud (1828-1890), **Odilon Redon** (1840-1916) painted microorganisms with human features.

For Richard Verdi, many paintings by **Paul Klee** (1879-1940) portray "the secretive world of microscopic life". Examples include "Pflanzlich-Seltsam" (Plant-like Strange) (1929), and "Vorhaben" (Intention) (1938), with its juxtaposed macrocosm and microcosm, external and internal worlds mediated by the human form.

Introduction

The Decorative Photo-Micrographs (1931) by **Laure Albin-Guillot** (1879-1962) are an especially noteworthy antecedent in terms of their attention to pattern and abstraction of vegetal forms through the amplifier of the microscope, thereby breaking down the barriers between science and the visual arts.

In the early decades of the 1900s, the artistic obsession with science and new technology led Italian writer **F. T. Marinetti** (1876-1944) and the Futurists to support Fascism, while other artists such as the renowned Mexican muralist **Diego Rivera** (1886-1957) embraced utopian communist ideals, especially in Rivera's "Man, Controller of the Universe" with its depiction of a series of plants, a microscope, and cellular life at the centre of the painting.

Regarding **Karl Blossfeldt** (1865-1932), whose close-up photographs of plants in *Urformen Der Kunst* (1929) redefine originary forms of nature as abstraction,

above: Wassily Kandinsky, "Striped" (1934)
opposite top: Joan Miró, "Carnival of Harlequin" (1924)
opposite left: Paul Klee, "Small World" (1914)
opposite middle: Wassily Kandinsky and Paul Klee, posing in jest as Goethe and Schiller, beach at Hendaye, France (1929)
opposite right: Odilon Redon, "Il y a peut-être une première humanité essayée dans la fleur" (1890)

14 Microcosms – Sacred Plants of the Americas

Introduction

Walter Benjamin wrote the following in 1928: "whether we accelerate the growth of a plant through time-lapse photography or show its form in forty-fold enlargement, in either case a geyser of new image-worlds hisses up at points in our existence where we would least have thought them possible".

Lázló Moholy-Nagy (1895-1946), the pioneer of Bauhaus biofunctionalism, is characterised by Oliver A. I. Botar as the "prototype of the progressive, avant-garde, techno-optimistic and media-optimistic artist". His works *The New Vision: From Material to Architecture* (1932), and *Vision in Motion* (1947) remain visually compelling and provocative. His flower photographs from the 1920s are particularly evocative abstractions based on organic forms.

Carl Strüwe (1898-1988) is the German author of *Formen des Mikrokosmos* (Forms of the Microcosmos) (1955), an astonishingly beautiful collection of 280 photographs taken through microscopes over a period of some three decades. Publicity materials for a solo exhibition at the Brooklyn Museum in 1949 affirm that Strüwe's microphotographs "often remind us of modern artists such as Klee or Kandinsky and yet they do not encroach upon the field of painting. Rather they suggest possible sources and explanations for modern abstract art, unearthing a whole world of beauty invisible to the naked eye". One of the featured works of the Strüwe exhibition at the Steven Kasher Gallery in New York in 2016 was "Archetype of Individuality" (1933).

György Kepes (1906-2001), the Hungarian artist, photographer, designer, and educator, who collaborated with Moholy-Nagy, is the author of the still highly-relevant *The New Landscape in Art and Science* (1956) as well as the *Vision + Value* series (1965-1972). Art theorist Charissa N. Terranova believes that Kepes's photographs recast "scientific utility as abstract art". She believes his work is best described as "vision exteriorised by technology". His photograms, for example, produced in Chicago from 1938-1942, were made without a camera by arranging natural objects directly on light-sensitive paper in a darkroom.

Additionally, one should not underestimate the influence of French philosopher **Henri Bergson** (1859-1941), whose concept *élan vital* from *Creative Evolution* (1907) was instrumental in defining how Bergsonian time links biological production and the generation of works of art.

As a boy, environmental philosopher **Michael Marder** was sent by Soviet doctors in Moscow on a trip south to Ukraine where the friendlier climate was to heal his illness. Instead, it put him in the invisible uncertain path of the radiation fallout from the Chernobyl nuclear plant disaster of April 1986. In *The Chernobyl Herbarium: Fragments of an Exploded Consciousness*, Marder collaborates with contemporary French visual artist **Anaïs Tondeur** to produce an overpowering and painful meditation on what Marder calls

top: Karl Blossfeldt, "Pumpkin Tendrils", *Urformen der Kunst* (1929)
above: László Moholy-Nagy, Flower Photogram (1926)

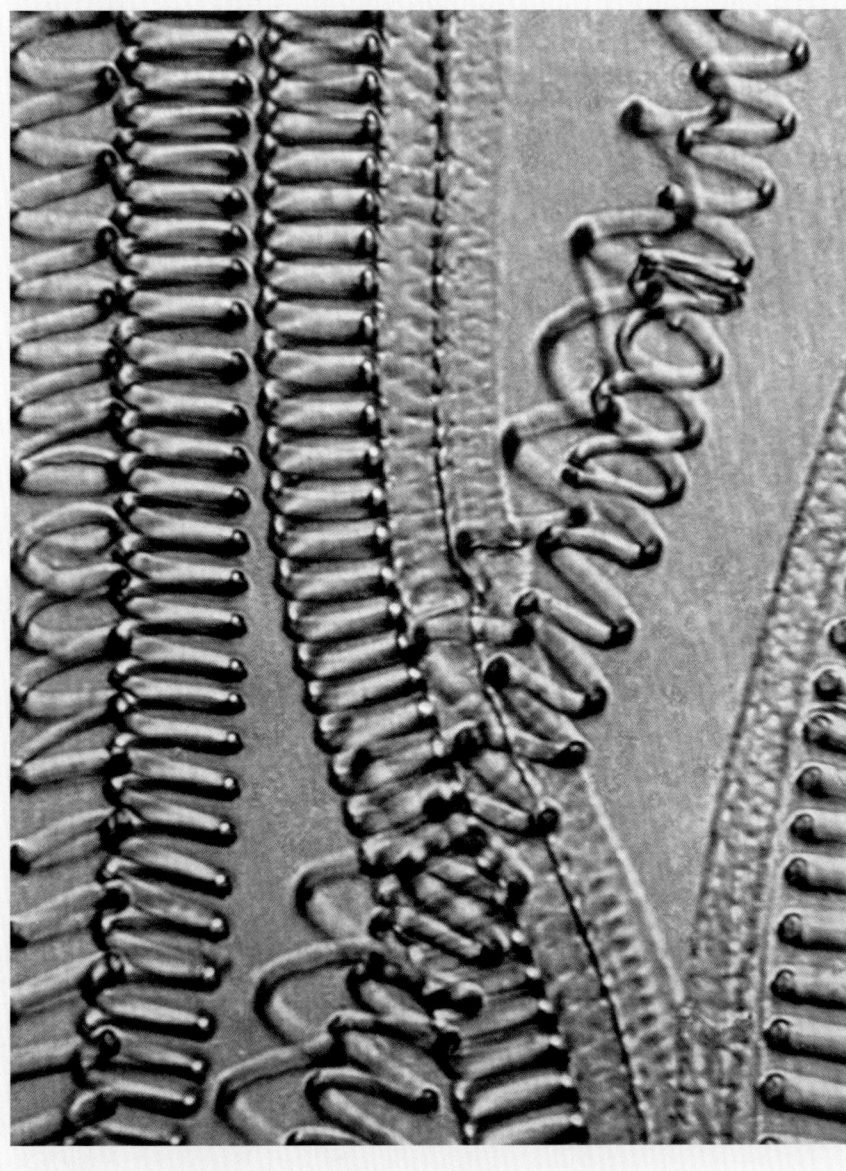 

above left: Carl Strüwe, Xylem, "Lindenholz" (Basswood), *Formen des Mikrokosmos* (1955)
above right: Carl Strüwe, Pollen, Peruvian Sunflower, *Formen des Mikrokosmos* (1955)

"life's vulnerability, amplified by the failure of reason to protect us on the hither side of the beautiful/sublime divide". Tondeur's works are photograms "created through the direct imprints of radioactive herbarium specimens, grown in the soil of "the exclusion zone" by Martin Hajduch of the Institute of Plant Genetics and Biotechnology at the Slovak Academy of Sciences and arranged on photosensitive paper". The authors hope that their collaboration cultivates a "more environmentally attuned way of living". *The Chernobyl Herbarium* (2016) is freely accessible from Open Humanities Press.

In the world of contemporary art, **Alexis Rockman**'s incorporation of microscopic images in his paintings is the basis of "Drop of Water" (2017) from his project "The Great Lakes Cycle" for which the artist, as he puts it, has created "a hybrid language that is natural history psychedelia". Finally, the magisterial micro-beauty in the Papadakis publications *Pollen*, *Seeds*, and *Fruit* (by **Rob Kesseler**, **Madeline Harley**, and **Wolfgang Stuppy**) is in a precedent-setting class of its own.

opposite: "Visión de un mundo místico" (2001) by Wixárika (Huichol) artist Santos Motoapohua de la Torre

THE SACRED AND THE SMALL
Ralph Metzner describes two common comparisons made by writers regarding experiences with psychoactive substances: "one is the *amplifier analogy*, according to which the drug functions as a non-specific amplifier of both inner and outer stimuli […] The other analogy is the *microscope metaphor*. It has been repeatedly said that psychedelics could play the same role in psychology as the microscope does in biology: opening up realms in the human mind to direct, repeatable, verifiable observation that have hitherto been largely hidden or inaccessible".

According to Hope MacLean, "the Huichol artist Alejandro López de la Torre […] told me that when we look into the world of the gods, it is as though we are looking through a telescope. The gods appear very tiny or far away. The same thing happens when shamans look into their mirrors. The gods are visible as small round images, just like images seen through the wrong end of a telescope". One example of this phenomenon is "Visión de un mundo místico" (Vision of a Mystical World) from the Museo Zacatecano de Arte Huichol by Santos Motoapohua de la Torre.

NAMING THE PLANTS
The digital images of plants in this book are identified by their scientific names (binomial nomenclature = genus + specific epithet), families, and also their common names in a wide variety of Indigenous languages, Spanish, and English.

Wade Davis has some interesting ideas regarding the process of naming and categorising plants based on his experience with the Indigenous peoples he has consulted on his many journeys into the Amazon:

"Wepe, like all the Waorani I met, turned out to be not only a keen observer but an exceptionally skilled naturalist. He recognised such conceptually complex phenomena as pollination and fruit dispersal, and he understood and could accurately predict animal behaviour. He could anticipate the flowering and fruiting cycles of all edible forest plants, list the preferred foods of most forest animals, and identify with precision the places where they slept.

It was not just the sophistication of his interpretations of biological relationships that impressed me; it was the way he classified the natural world. He often could not give you the name of a plant, for every part – roots, fruit, leaves, bark – had its own name. Nor could he simply label a fruit tree without listing all the animals and birds that depended on it.

His understanding of the forest precluded the narrow confines of nomenclature. Every useful plant had not only an identity but a story…".

LOOKING TO THE FUTURE
John C. Ryan, author of the groundbreaking study *Posthuman Plants: Rethinking the Vegetal through Culture, Art, and Poetry*, declares unequivocally: "the receiving of ethnobotanical good should be balanced by a giving back of good to the plants themselves, the environments in which they grow naturally, and the Indigenous people whose cultural heritage involves medical knowledge of the species. It is not enough to privilege cultivating healing plants as a solution to their disappearance in the wild. As species decline, the ecocultural knowledge systems associated with them become at risk…".

He continues, saying "a potential transformative outcome of ecodigital art is the changing of public perceptions and behaviours concerning nature and humanity's fractured relationship to plant life". To do this work effectively, Ryan calls for *inter-disciplinarity*: "is an ecodigital practitioner an environ-mentalist, artist, poet, scientist, engineer, conservator, botanist, or all of the above?".

In closing, it is urgently important to highlight what Jonathan Ott says in his magnum opus *Pharmacotheon*: "I firmly believe that contemporary spiritual use of entheogenic drugs is one of humankind's brightest hopes for overcoming the ecological crisis with which we threaten the biosphere and jeopardise our own survival, for *Homo sapiens* is close to the head of the list of endangered species".

The plant-teachers in this book, *Microcosms – Sacred Plants of the Americas*, respected, seen together, and magnified in aesthetically-innovative ways previously unknown, can lead the way toward a shift in consciousness.

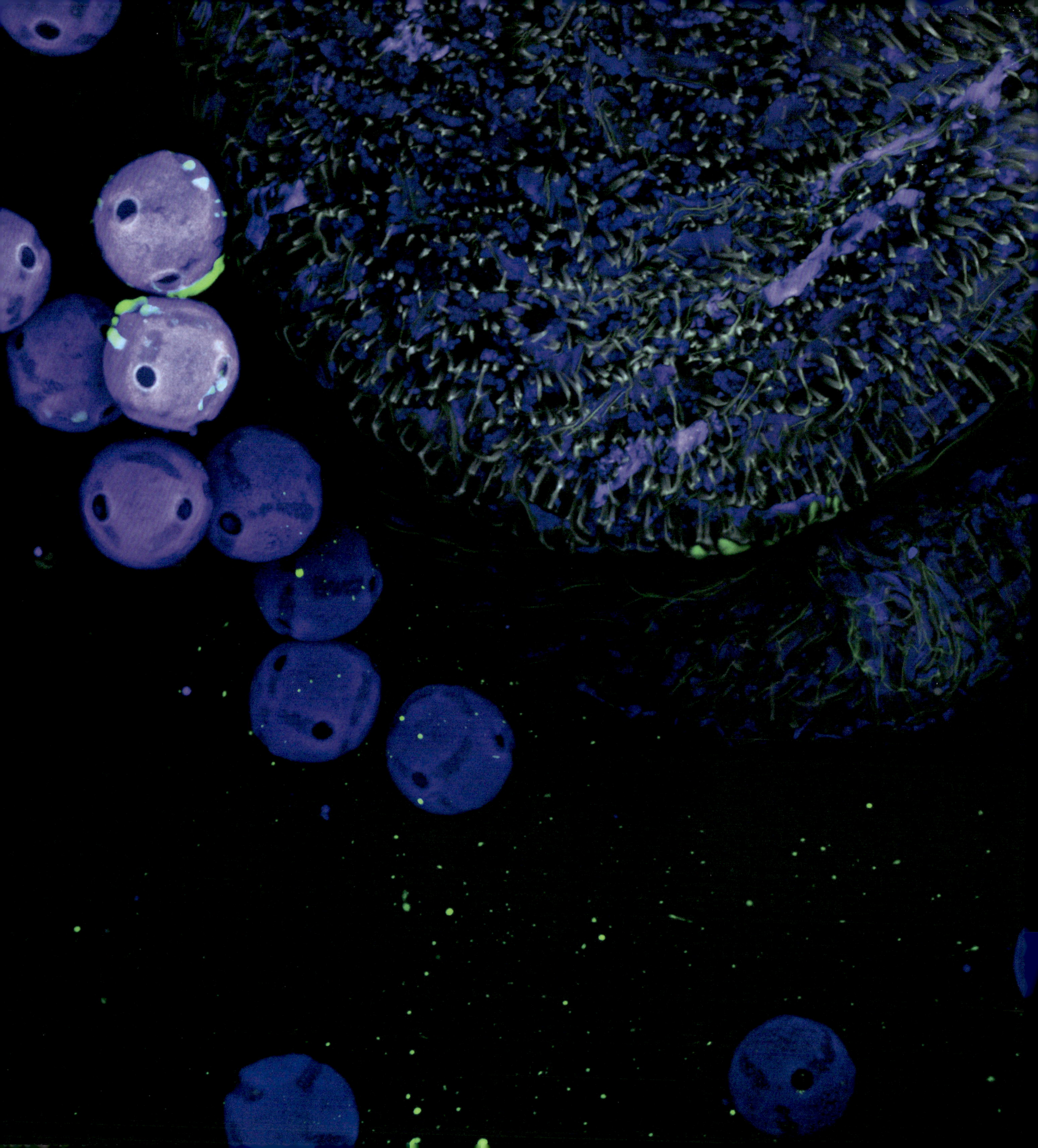

Microcosms
The Inner Worlds of Sacred Plants
Through the Lens of Science, Art, & Ancestral Knowledge

Steven F. White

This book is a tribute to Indigenous wisdom keepers such as Salvador Chindoy, Robin Wall Kimmerer, Davi Kopenawa, Fernando Payaguaje, María Sabina, Enrique Salmón, and many others, who have been stewards of the ethnobotanical knowledge that has been transmitted over many successive generations. The images and theoretical framework that constitute *Microcosms – Sacred Plants of the Americas* exemplify innovative art-science projects that, according to Joanna Page in *Decolonising Science in Latin American Art*, "increase our understanding of other species by promoting an aesthetic and affective engagement with new scientific findings, or deploy scientific techniques for objectives other than those of predicting, controlling or commodifying the natural world".[1] Microscopy Specialist Jill Pflugheber and I, a dedicated "Plant Ally", created the contents of this homage over nearly a decade. By questioning the kind of rational scientific Western model that leads to the commercialisation of sacred plant medicines as well as the erasure of Indigenous Peoples, *Microcosms* seeks to define an ethical responsibility to understand how Scientific Ecological Knowledge (SEK) and Traditional Ecological Knowledge (TEK) can grow and thrive together, potentiating each other more effectively. For Robin Wall Kimmerer, these two knowledge systems need to exist in a sovereign, autonomous manner based on mutual respect. This is the way to resist cultural imperialism, democratise science to increase its accessibility, and begin to free science from its determination to commodify intellectual property with its unacknowledged and unattributed Indigenous origins in order to exploit it for private financial gain.[2] In order to accomplish this, *Microcosms* offers an exciting journey into the inner worlds of plants through the lens of science, art, and ancestral knowledge. In his pioneering study "On Being Called by Plants: Phytopoetics and the Phytosphere",

Banisteriopsis caapi (ayahuasca / yagé); confocal image

50 μm

John C. Ryan divides what he calls the phytosphere into three domains: "in contrast to rhisospheric poetry of the root-soil interface and phyllospheric poetry of the leaves, endospheric poetry directs the human sensorium to the interior of plants. The endosphere is a site of micro-organismic transactions as well as plant communication via chemical compounds, electrical signals, and other means".[3] For Ryan, *Microcosms* is a work of "endospheric visualisation" in that it "forges an optical language for generating engrossing depictions of the innermost topographies of sacred flora".[4]

Although throughout our project we proceeded under the assumption that all life deserves veneration, the work we did was informed by a definition of sacredness along the lines of what Gary Paul Nabhan describes in a recent *Bioneers* interview: "I define a sacred plant as any plant that's used in personal or collective rituals, ceremonies, seasonal rights, or sacramental traditions, so it includes not only psychedelic plants, many of which are sacred in their original traditional settings, but many other species [...] The sacredness of a plant has to do with the complex interactions surrounding it, not in matter itself".[5] It would be presumptuous and entirely inaccurate to affirm that the more than fifty species that comprise *Microcosms* are *the* sacred plants of the American continent. But the book offers an engaging point of departure with *some* of the most interesting plant entities you will ever meet. We wanted to appreciate their histories and share some of their stories, narratives linked to lands inhabited by Native peoples who had their own names for this vast region, and longstanding spiritual relations with their plant relatives. As Keith J. Williams writes, "since pre-colonial times, Indigenous peoples from Turtle Island to Abya Yala have considered sacred and visionary plants as living beings, with which it is possible to communicate through ritual and ceremonial languages, and, according to Indigenous ontologies, these sacred plants are not isolated from the territory".[6]

Microcosms reveals certain formal botanical structures, confined here for the sake of expediency to **stomata, trichomes, xylem,** and **pollen**. These bioforms are clearly visible in the confocal plant-art images constituted by sacred beings that grow in the most diverse American landscapes imaginable, from deserts to rain forests. To appreciate and analyse these sites of microcosmic aesthetic contemplation, one needs to take into account without fail the biologically-complex habitats of these plants, how these interconnected ecosystems are threatened, and what needs to be done immediately to protect them. I concur with the primary objective of Charissa N. Terranova and Meredith Tromble, coeditors of the *Routledge Companion to Biology in Art and Architecture* when they affirm: "our goal is to show that art-and-science endeavours are not bound to any one way of being: they are indicative neither of the violent takedown of a reified and mystified science nor of a passive instrumentalised relationship in which art is the servant of science".[7] Maura C. Flannery, author of *In the Herbarium: the Hidden World of Collecting and Preserving Plants*, maintains that the aesthetic appreciation of plants facilitates a deeper understanding of the physical world and the urgent need to preserve it. Her characterisation of herbaria also applies to the ecodigital garden of plant-art that constitutes *Microcosms*: "herbarium collections and individual specimens can be treated as works of art as well as of science in terms of how the plants were arranged and embellished. This aesthetic aspect is also pivotal to the work of artists now studying plant collections and using them as inspiration for contemporary work".[8]

Stomata, for example, open and close pores on the surface of leaves and stems to allow plants to take in carbon dioxide and release oxygen as a byproduct of photosynthesis. The apertures are surrounded by bean-shaped parenchymatic cells (guard cells), which, through astonishingly complex signalling networks, control the size of the openings in order to regulate the gaseous exchange and control transpiration. Environmental stimuli that influence this process include relative humidity, carbon dioxide concentration, light intensity, and temperature. Stomata should not be considered passive portals of entry to the plant since current research has demonstrated stomata-based defences against destructive bacteria. We have highlighted in these pages the abundantly-visible stomata of *Brunfelsia*, and wonder if it

Brugmansia insignis (angel's trumpet); confocal image

20 μm

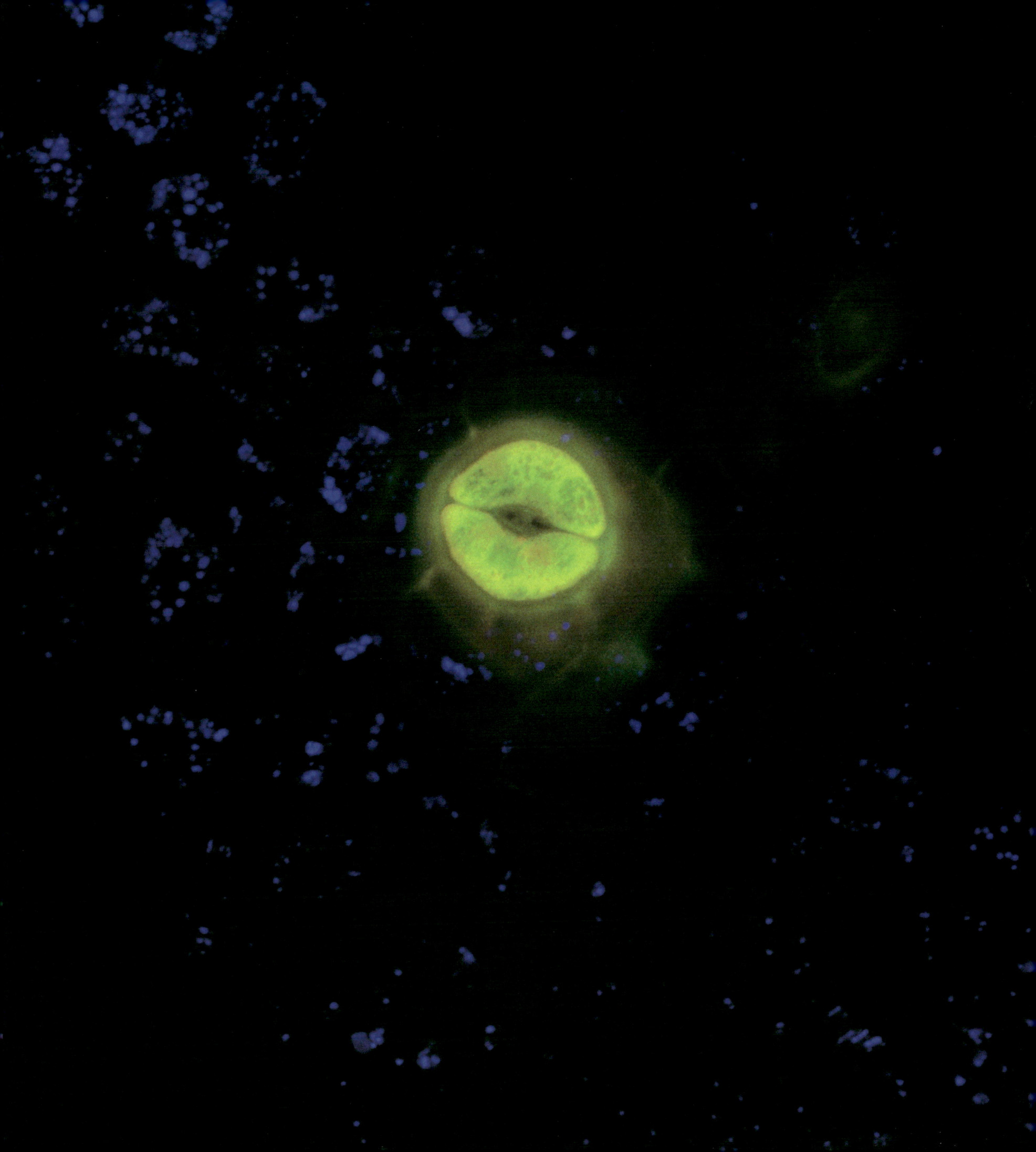

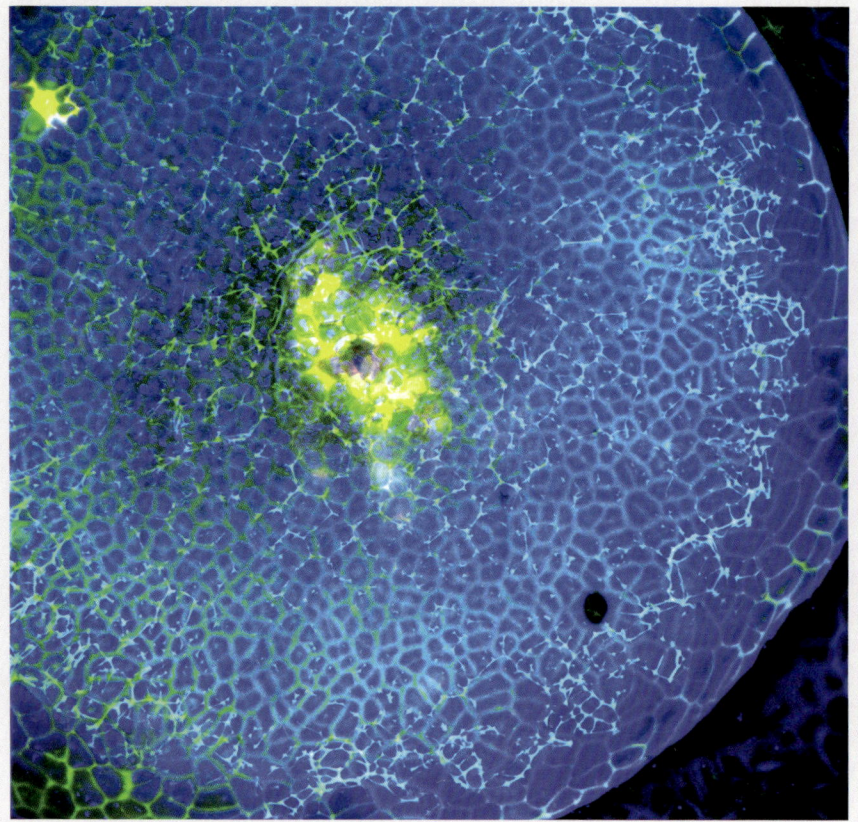 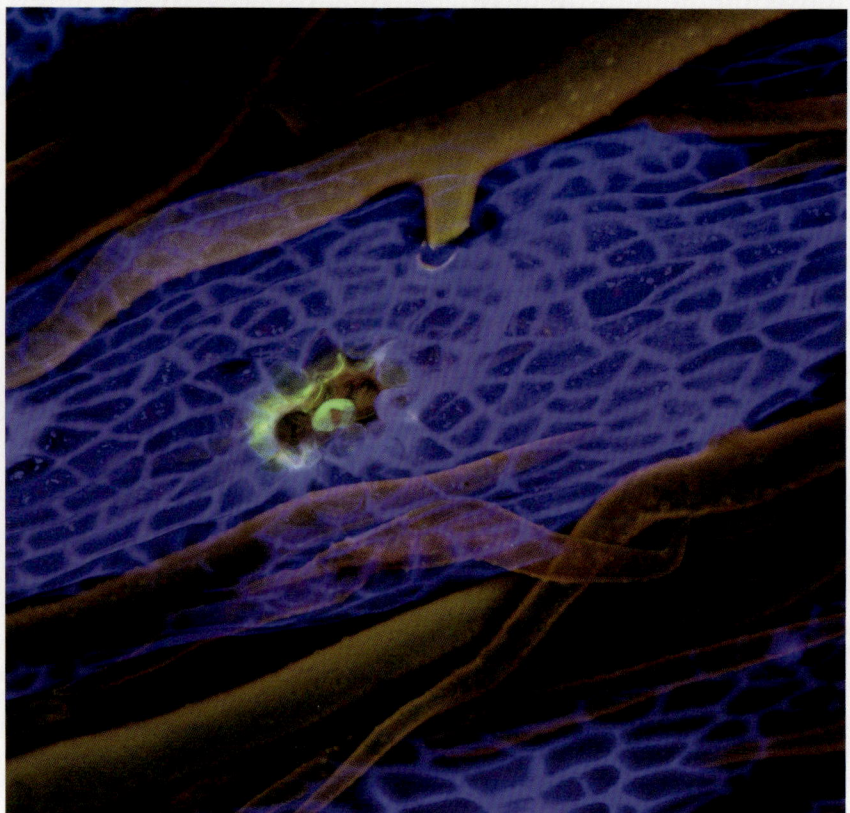

would be possible for one to imagine breathing with this art in its living form. The stomata of these varied species could also reveal a different layer of time, since fossil plants with relatively fewer stomata indicate higher carbon dioxide levels and, therefore, plants that lived in times of high global temperatures. When some future living thing traverses the fossils of plants from our Anthropocene epoch, where will we be as a species? Extinct?

My contemplation of the stomata that appear in a wide variety of confocal images of plant species included in *Microcosms* coincided with an aspect of Native American philosophy that came to orient and underlie our project. In *Iwígara, the Kinship of Plants and People: American Indian Ethnobotanical Traditions and Science*, Enrique Salmón addresses what he calls "kincentric ecology", and describes the simple but powerful notion that life is interconnected and shares the same breath.[9] All species have an equal right to inhabit and flourish on Earth. These ideas resonated deeply in my mind as I read Emanuele Coccia's *The Life of Plants: A Metaphysics of Mixture*. Breathing and shared breath are at the very heart of his highly-poetic study: "thanks to plants, the Earth definitively became the metaphysical space of breath. The first to colonise and make the Earth inhabitable were the organisms capable of photosynthesis: the first living beings that were wholly terrestrial are the greatest transformers of the atmosphere".[10] "In breath", asserts the author, "the first and the most trivial and unconscious act of life for a huge number of organisms, we depend on the lives of others".[11]

How is it that an exquisite stained-glass beauty of compartmentalised fluorescence in *Psychotria viridis* and *Diplopterys cabrerana* characterises these two plants that are the DMT-source in the sacred beverage ayahuasca/yagé? The repeating patterns of pavement cells that can be observed in the confocal images of *Anadenanthera colubrina* and *Desfontainia spinosa* create a natural jigsaw puzzle of life's intimately interlocking pieces that are a means to behold an

above left and right: *Banisteriopsis caapi* (ayahuasca / yagé); confocal images   50 μm
opposite: *Datura innoxia* (toloache); confocal image

50 μm

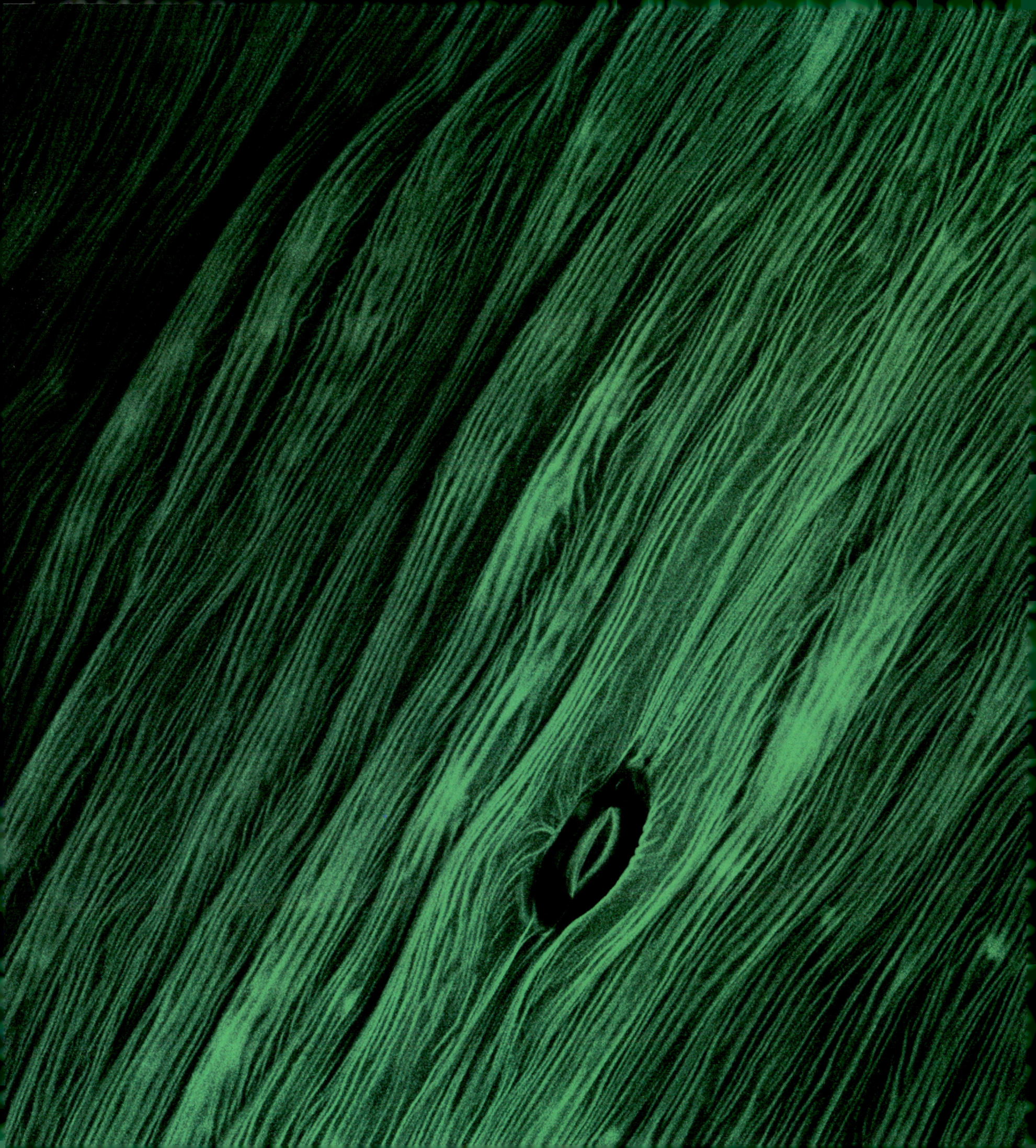

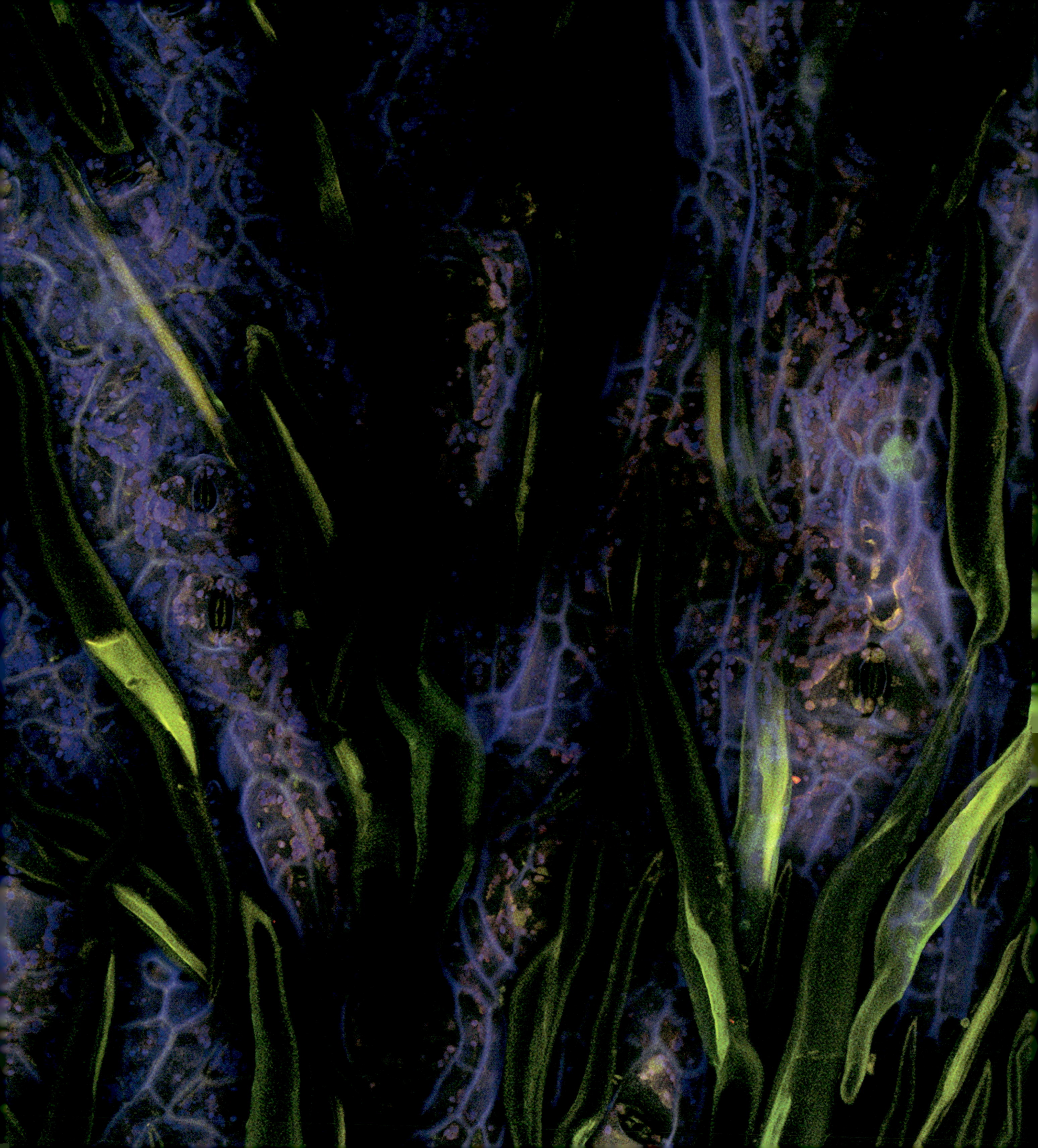

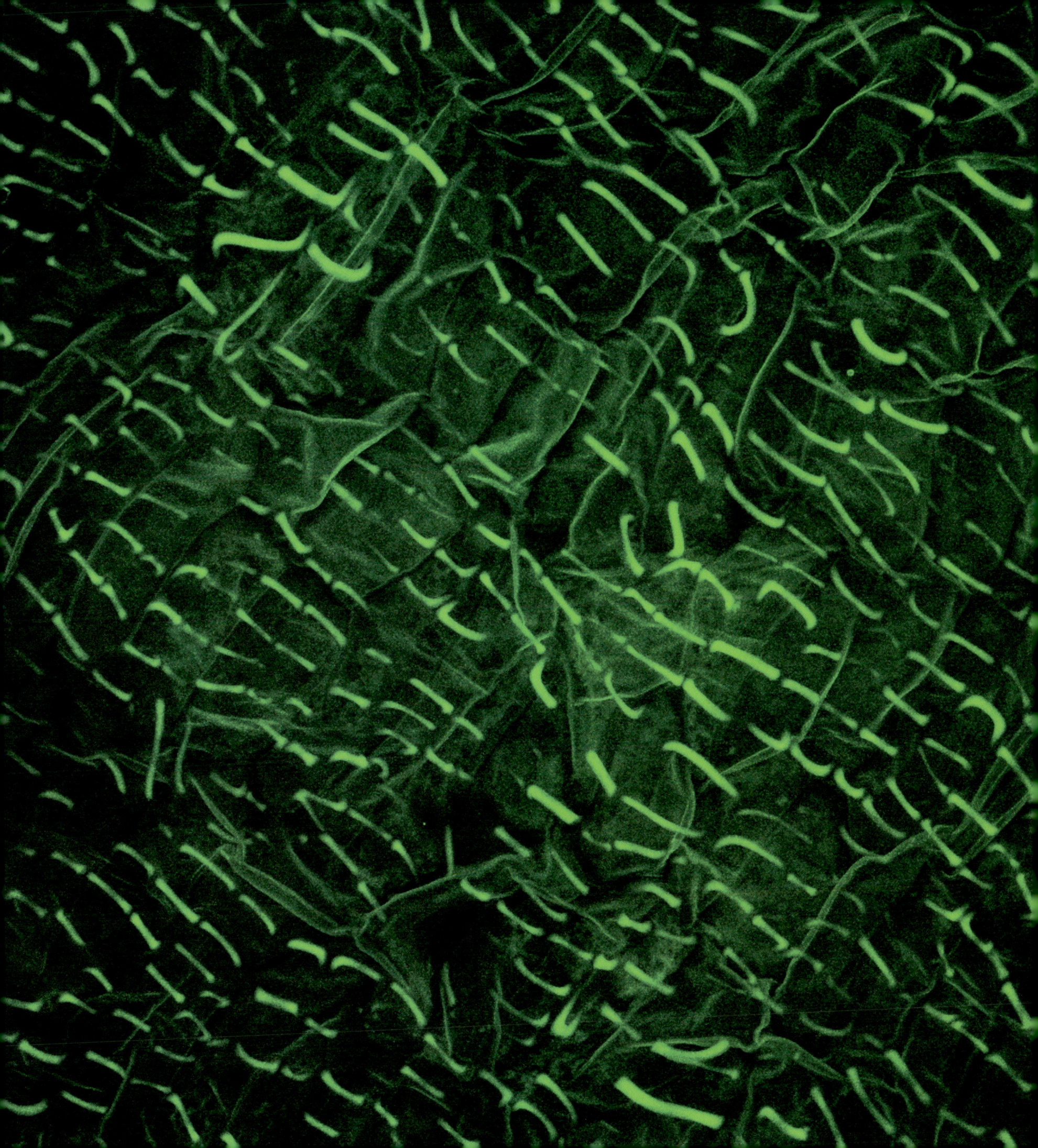

evanescent beauty in a frozen image captured from living material. Perhaps the confocal images themselves could be conceived as *visual alkaloids*, the equivalent of the stimulant defined by Jonathan Ott as "nitrogen-containing organic compounds which represent the pharmacologically-active principles of many plants",[12] including a majority of those that are gathered in the pages of *Microcosms – Sacred Plants of the Americas*. Why couldn't these images be contemplated as potentially therapeutic pathways to promoting altered states of consciousness? If each species were to be considered an artist in and of itself, the vegetal patterns and structures of the "self-portraits" might be viewed in light of J. P. Hodin's idea of "the painter's handwriting": *écriture, style, touche*.[13] These images are examples of the "writing" of plants, a manifestation of their personality, a clue to their traditional affiliations, inimitable in its subtlety, the touch not only of cycles of growth in an individual plant but also of evolution. Underlying this distinctive style are the elements of great works of art: structure, weight, density, motion.

The **trichomes** (from its Greek etymological origin meaning "hair"), too, are so utterly distinctive, as in the case of the principal plant used in the preparation of the sacred beverage ayahuasca: *Banisteriopsis caapi*. When I first saw the way these T-shaped trichomes are attached, understanding that this is a trait of the Malpighiaceae family, I was reminded of the tail of a whale, after it has breached and dives again into unimaginable depths. And you, reading this book as you snack on a bar of your favourite chocolate, will surely be transformed by the sublime five-fingered stellate non-glandular trichome of *Theobroma cacao*! One of the primary functions of trichomes is to provide the plant physical and chemical protection from microbes and insects. Our image of *Ipomoea corymbosa* (Ololiuhqui) possesses dramatic activated sentinels guarding precious and strategic inner worlds. Or perhaps trichomes function as antennae, as Zoë Schlanger suggests in *The Light Eaters: How the Unseen World of Plant Intelligence Offers a New Understanding of Life on Earth*: "researchers have already found that trichomes allow plants to sense the footsteps of moths and caterpillars, and mount defenses in response; trichomes are clearly exquisitely sensitive organs".[14]

Another basic plant-form worthy of contemplation is **xylem**. Xylem vessels are vascular bundles in plants that serve to transport water and minerals with a unidirectional flow from roots to leaves. Located at the centre of the plant, they are composed primarily of dead cells to facilitate greater capacity to carry water. With walls reinforced with lignin, xylem vessels have two primary forms, both of which are aesthetically inspiring and can be found in these confocal images. Strangely enough, the coiled xylem is particularly impressive in the confocal image of the formidable Brugmansia known as "culebra" (snake). In annular vessels, the lignin forms a pattern of equidistant circular rings. With regard to spiral vessels, the lignin resembles a helix or coil. One need not be a botanical expert to begin to appreciate the dynamic vitality and beauty of these forms that enable systems requiring cohesion and adhesion to function. The confocal images with their stunning depictions of xylem instill a deep appreciation of the flow of water that ensures the survival of all living things.

The aesthetic marvels of fundamental observable *microcosmic phytoformalism* would not be complete without **pollen**, which delivers male gametes to an ovule of a compatible flower to fertilise an egg that then becomes a seed. It was not always possible for us to image pollen and its astonishing variable beauty, but we were very impressed with the results when we did get the chance to work with flowers from the sacred plants. Pollen grains from Angiosperms range in size from 25-250 microns in diameter. The pollen grains in *Microcosms* are 50 microns or smaller. The outer wall of the pollen grain, called the exine, usually has ornately sculptured three-dimensional designs that are species-specific and is composed of a highly-resistant biopolymer. How tough is it? As the scientists Dobritsa and Wang maintain in their research: "the extraordinary stability of sporopollenin accounts for exine preservation for hundreds of millions of years in fossil material and allows palaeontologists to address questions related to past states of climates and

page 26: *Banisteriopsis caapi* (ayahuasca / yagé); confocal image **page 27:** *Datura innoxia* (toloache); confocal image
opposite: *Lophophora williamsii* (peyote); confocal image

50 μm

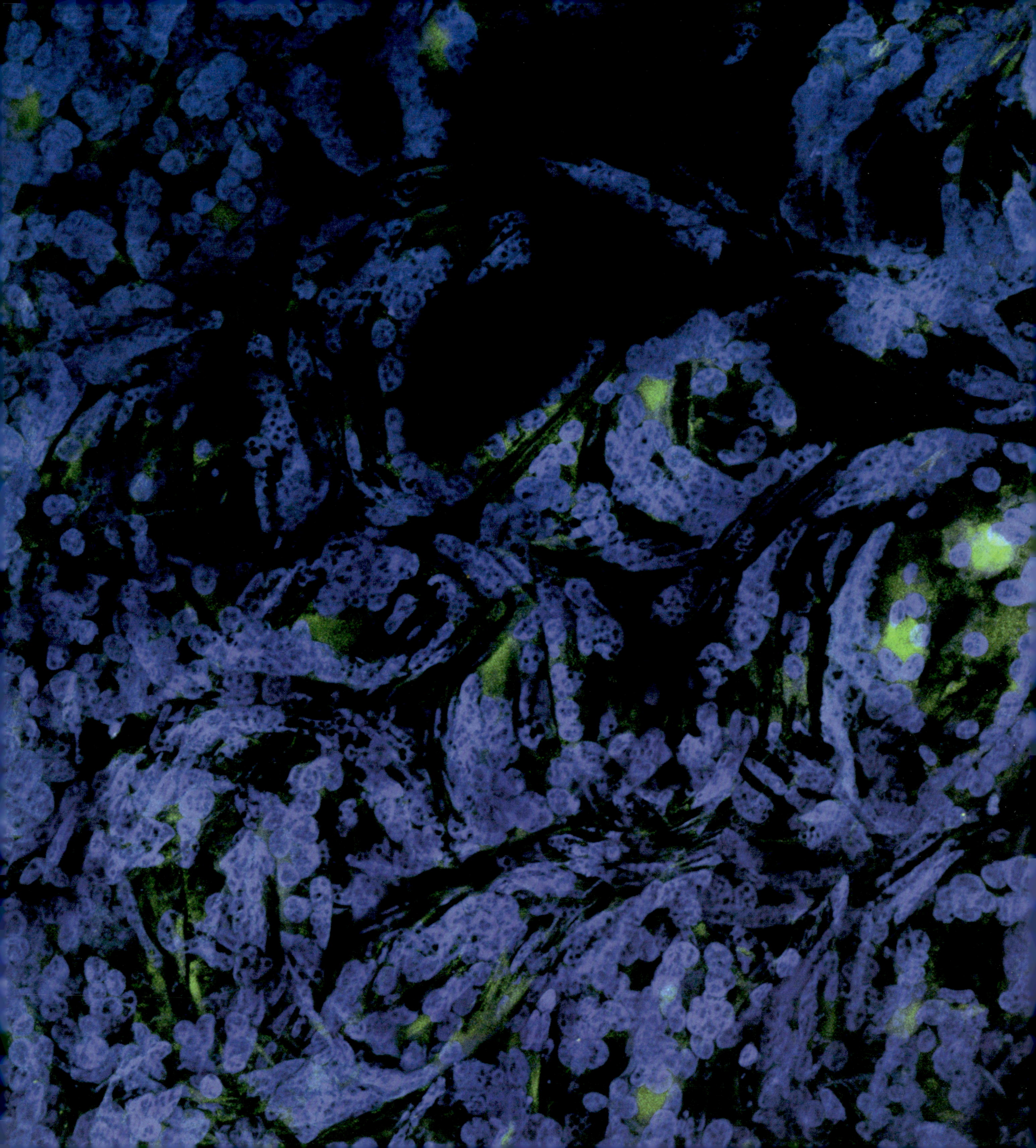

vegetation and determine evolutionary relationships between plants".[15] The geologic history of terrestrial plant life can be studied by means of the virtually indestructible exines preserved in sediment. Pollen is art that creates itself to last! Pollen (microscopic, gloriously sculpted, and far beyond any human capacity to remain intact) as one of our sites of contemplation, then, obliges us to consider co-existing plant-time-scales. In itself, pollen is another door of perception to an understanding of a physical world (as well as our place in it) that is both immediate (in terms of plant reproduction) and evolving over vast periods of time.

Microcosms is a widely-accessible repository of plant-teachers and the biocultural history that accompanies them. We consider this an ideal collection of allies who can help us undertake what needs to be done in order to halt environmental degradation and climate change. These plants, among my very favourites as an anthologist of vegetal lives, cannot *not* be together. They interact with each other. Through *Microcosms*, we can share their inner lives and assimilate them, getting to know ourselves more fully at the same time. These digital images are a way of re-understanding, undoing our indifference and ignorance, coming to appreciate the sacred plants, and taking this knowledge with us to inhabit the physical world more responsibly. Rob Kesseler, in "Pixillated Pollen" from the Papadakis book *Pollen – the Hidden Sexuality of Flowers*, writes that "the digital world we occupy provides a fertile breeding ground for art-science initiatives".[16] He proposes "transforming an exotic fusion of scientific knowledge and artistic interpretation into a personal *phytopia*".[17] I agree wholeheartedly. Ultimately, what I was looking for, based on my studies of art history, from palaeolithic drawings that I saw firsthand during multiple visits to the caves in northern Spain to the most cutting-edge work in galleries and museums of major urban centres in the United States, Europe, and Latin America over the decades, were images with the power to root themselves in memory. I found them in the magnified inner lives of these sacred plants, and they will not be forgotten. Is it evident by now how, in these confocal images, the sacred plants of the Americas, with their distinct forms, colours, juxtaposed textures and composition, aspire to be art, creating an aesthetic proposal that has not been seen before and cannot be repeated in precisely the same way again?

I fervently hope that the extended, ultra-sensory worlds revealed by *Microcosms – Sacred Plants of the Americas* will contribute in some small way to new forms of egalitarian thought and laws that provide the earth, its waters, and its atmosphere protection from the violent human exploitation that must become as unthinkable and impermissible as slavery. The physical world has many powerful human enemies who will stop at nothing to perpetuate the same models of business as usual with their extractive systems that will eliminate finite natural resources in the interest of accumulating capital. It should be said that the formidable array of psychoactive plants of power throughout the Americas was always used by the Amerindian population as a necessary tool for maintaining social cohesion and shared cultural values with regard to the environment, and also as an effective technology to resist, to wage war, and to defeat their adversaries.

The tragic history of the Americas during the colonial period (and beyond) in many ways can be understood as the violent assault by European invaders on the sacred plants and the Indigenous leaders who consulted these plant-teachers as a way of benefiting their societies and learning how to live in greater (though hardly perfect) equilibrium with the natural world. It is not my intention to idealise or romanticise Indigenous cultures that clearly left their human mark on the landscapes they inhabited and shaped. To facilitate the looting of the continent, the foreigners first needed to destroy the gods of their new slaves, as well as the food and religious ceremonies that sustained them. I've dreamt of Spanish soldiers furiously chopping down the red plumes of amaranth, bursting into healing rituals to confiscate the *cohoba* of a Taíno shaman, destroying the mushroom-children of one of María Sabina's ancestors, torturing Indigenous prisoners to extract ever more information about the vegetal lives that are such an integral part of Amerindian lives, banning the use

Trichocereus macrogonus var. *pachanoi* (huachuma / San Pedro cactus); confocal image

50 μm

of ayahuasca upon pain of condemnation to infernal realms, burning the botanical knowledge preserved in codices, and demonising the systems of cosmic plant-patterns that oriented pre-Hispanic existence. These official impositions, though challenged in the contemporary world, largely remain intact, and seriously endanger anyone who denounces and breaks these unjust and hypocritical laws. I am remembering, too, my journey to the Ecuadorian Amazon in the late 1970s when I visited a Cofán community and saw firsthand how Evangelicals from the United States had decimated ancestral traditions related to the use of sacred plants. The only thing there that saddened me more at that time was the recently-finished 500km pipeline snaking from Lago Agrio over the Andes toward the Balao Pacific port, carrying the oil that ruined so many lives and ecosystems. Christianity and extractivist economies have been operating together with devastating consequences for Indigenous peoples in the Americas for centuries.

There is a reason why Amerindian religious systems deified Nature itself. If we doubt the efficaciousness of cosmogonies that contribute to the survival of *all* species, unfortunately, we are lost. In the *Popol vuh*, the Hero twins, Hunter and Jaguar Deer, descend into the underworld of Xibalba. They are only able to triumph over the Lords of Sickness and Death when they form close alliances with a wide variety of non-human species. In one of my favourite parts of the narrative, fireflies save the twins from execution by pretending to be the lit ends of cigars of sacred tobacco that were not to be consumed in their entirety during a long night of trials. Thank you, *ch'umk'ak'* ("firefly" in Quiché Maya)! In his extraordinarily gifted English translation of the *Popol vuh*, Dennis Tedlock speaks of how the "Council Book" itself was considered by the lords of Quiché "an *ilb'al*, 'a seeing instrument' or a 'place to see'; with this they could know distant or future events".[18] Curiously, as Tedlock also explains, "today *ilb'al* [...] refers to crystals used for gazing by diviners and to eyeglasses, binoculars, and telescopes".[19] Could this be extended to include the confocal microscope as yet another instrument for augmenting the limits of human vision? And perhaps *Microcosms*, as a gathering of vegetal lives and their stories in relation to humans, might also be considered a privileged and strategic vantage point, an aesthetic site that reveals better ways to proceed into a future that is bound to be grim and dominated by what most concerned the Maya seers: extreme weather conditions, famine, and war.

By way of a conclusion to this essay produced in a small, remote town in rural upstate New York on land taken from the Mohawk Nation, I feel compelled, at least initially, to resist the obligatory optimistic note to express instead my shame and indignation about the ways that we, as humans, have misunderstood, mistreated, exploited, and demonstrated an utter lack of respect for plants. As Michael Marder affirms in his inspiring study *Plant-Thinking: A Philosophy of Vegetal Life*, for humanity in general, plants "have populated the margin of the margin, the zone of absolute obscurity on the radars of our conceptualities",[20] and that somehow we feel we have the right to operate on the assumption that "vegetal beings [are] unconditionally available for unlimited use and exploitation".[21] Obviously, the goal of *Microcosms – Sacred Plants of the Americas* with its ecodigital repository of biocultural heritage is not to turn these confocal images of more than fifty sacred plants from the Americas into static artefacts detached both from their ecosystems and from the Indigenous peoples whose ancestral knowledge has provided the most intimate, immediate understanding of *who* these vegetal entities are and what they demand from us in their role as emissaries from the natural world. On the contrary, I hope that this book becomes a platform for new aesthetic experiences through technology, a site of resistance to humanity's predominant utilitarian interest in plants, a means of denouncing the abuses that are producing a mass extinction of plant species, and a call to urgent, empathic, morally-based activism as conservators, creators, and informed citizens against the political and economic systems that are so irrevocably harmful to the environment.

Banisteriopsis caapi (ayahuasca / yagé); confocal image

50 μm

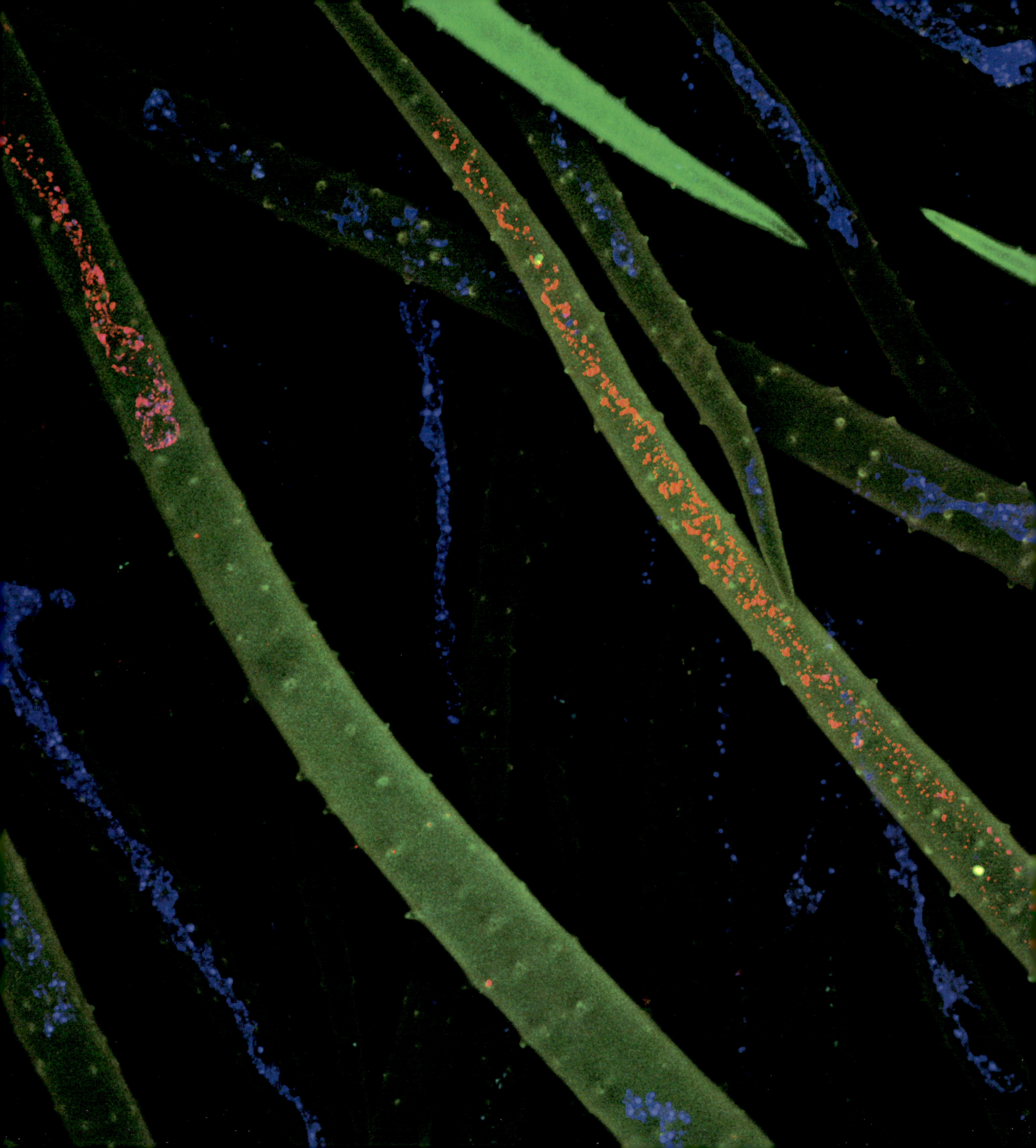

Malpighiaceae

Alicia anisopetala

Black ayahuasca, thunder ayahuasca

This woody South American liana bears a strong resemblance to Banisteriopsis caapi. Like most members of the neotropical Malpighiaceae family, it relies on oil-collecting bees for pollination. Although the plant is not psychoactive itself, it does fulfill an important ceremonial role as a purgative.

As Glenn H. Shepard, Jr. has written, "though a fair amount is now known about *how* psychoactive plants and compounds produce their peculiar effects on the human mind, it is still largely a mystery as to *why* certain plants produce such compounds".

In other words, why do some 100 plants from among perhaps half a million different plant species make these substances that can potentiate profound effects on humanity's consciousness of our destructive (or even its opposite more egalitarian) relationship with the natural world?

Does it indicate some kind of mutually beneficial co-evolution? Schultes and Hofmann call this "one of the unsolved riddles of nature".

The phytochemistry and ethnobotanical history of the rare plant *Alicia anisopetala* (from the same family as *Banisteriopsis caapi* and *Diplopterys cabrerana*, and often called "black" ayahuasca) are becoming clearer, though certainly warrant further research. A 2024 study conducted by a team of Australian scientists headed by Jonathan Tran used ultra-high performance liquid chromatography (UHPLC) coupled to mass spectrometry (MS) to search for the following six psychoactive compounds in *Alicia anisopetala*: tryptamine, N,N-dimethyltryptamine (DMT), 5-methoxy-N,N-dimethyltryptamine

Alicia anisopetala (black ayahuasca); confocal image

50 μm

(5-MeO-DMT), tetrahydroharmine (THH), harmaline, and harmine. What did they find? "No psychedelic alkaloids of interest or tryptamine were detected in our samples". Changing the research parameters, however, to focus on terpenes and other compounds of this plant could facilitate an understanding of an "entourage effect" in terms of how *Alicia anisopetala* is combined with other plants that *are* psychoactive. Does one of the confocal images included here capture a "blue" terpene emission from a trichome?

"Neil Logan affirms that Alicia anisopetala is also known as purgahuasca, and is used as a pre-ceremonial cleanse, an essential part of how these visionary plants are used efficaciously by Indigenous peoples."

above left and right: *Alicia anisopetala* (black ayahuasca); plant
opposite: *Alicia anisopetala* (black ayahuasca); confocal image

50 µm

Amaranthaceae
Amaranthus cruentus

Alegría, amaranth, bledo, huautli

This annual herbaceous plant with its telltale red plume of flowers grows in the seasonally dry tropical biome of Mesoamerica, and has been cultivated since ancient times for grain. Because it tolerates climatic adversity, its tiny reddish-white seeds are becoming an ever more viable alternative source of nutrition.

Ricardo Ortiz describes not only the enormous importance of amaranth as a basic food source in the pre-Columbian era, but also how it was used as a sacred plant in rituals dedicated to the Aztec (Mexica) war god Huitzilopochtli. The seed was an integral part of the festivals in which human sacrifices were made. Consequently, the Spanish missionaries took it upon themselves to abolish these religious ceremonies. Cortés ordered his soldiers to eradicate the amaranth plants that produce the seed, and they searched the countryside to destroy the distinctive red plumes of the sacred plant.

According to Ortiz, the Indians considered the seeds to have a mystical, transcendent power, as they were used to make a sacrament for honouring Huitzilopochtli that was taken in communion by adults and children, men and women alike. The sacrament was received with reverence, fear, and joy, because participants believed they were partaking of the flesh and bones of god.

Ortiz also rightly points out that "the Spaniards must have known that by eliminating the cultivation of *huautli*, the Indigenous people were being deprived of physical and

Amaranthus cruentus (amaranth); confocal image

100 μm

spiritual nourishment, which made it easier to subjugate them [to Spanish rule]".

In the more recent past, certain amaranth species cultivated in rural areas of Guatemala by Indigenous groups were nearly brought to extinction by the scorched-earth policies favoured by those US military advisors working with right-wing military governments in Central America in the 1980s in their wars against insurgent guerrilla movements.

According to Beilin and Suryanarayanan, South American eco-activists currently have adopted the radical strategy of using mud "bombs" containing glyphosate-resistant *Amaranthus palmeri* seeds to sabotage the fields of monoculture crops such as soybeans.

The amaranth "weeds" – whose cereal is, in fact, edible, – choke the ecologically-destructive plants grown primarily for export, whilst the amaranth plants themselves prove virtually impossible to eradicate.

Soriano-García and Aguirre-Díaz, in their overview "Nutritional Functional Value and Therapeutic Utilisation of Amaranth", recognise *Amaranthus* as an ancient and highly-nutritious New World crop consumed by the Aztec, Maya, and Inca civilisations. The researchers also document how, in contemporary usage, amaranth aids with "antihypertensive, antioxidant, antithrombotic, and antiproliferative biological activities".

A team of South African scientists led by Olusanya N. Ruth, in addition to pointing out the value of amaranth as a drought-tolerant plant with a "massive potential of curbing food-related problems", also indicate that, despite being considered a superfood, amaranth has been neglected and stigmatised as a "food plant for the poor". The article documents the numerous neutraceutical and healing properties of amaranth in its use throughout Africa, Asia, and the Americas. The researchers believe that more educational measures must be taken in order to fully communicate the enormous benefits of this plant.

Ukrainian researchers headed by O. L. Chulak focus their study on how amaranth oil can be used "to significantly slow down the processes of vascular hardening, and therefore reduce the possibility of developing a heart attack and stroke". The scientists make the following recommendation for the use of amaranth oil: for "prevention and treatment of cardiovascular diseases: coronary heart disease, myocarditis – take 30 minutes before meals, 1 teaspoon 2 times a day during complex treatment or 400-500ml per year for preventive purposes".

top and above: *Amaranthus cruentus* (amaranth); plant
opposite: *Amaranthus cruentus* (amaranth); confocal image

50 μm

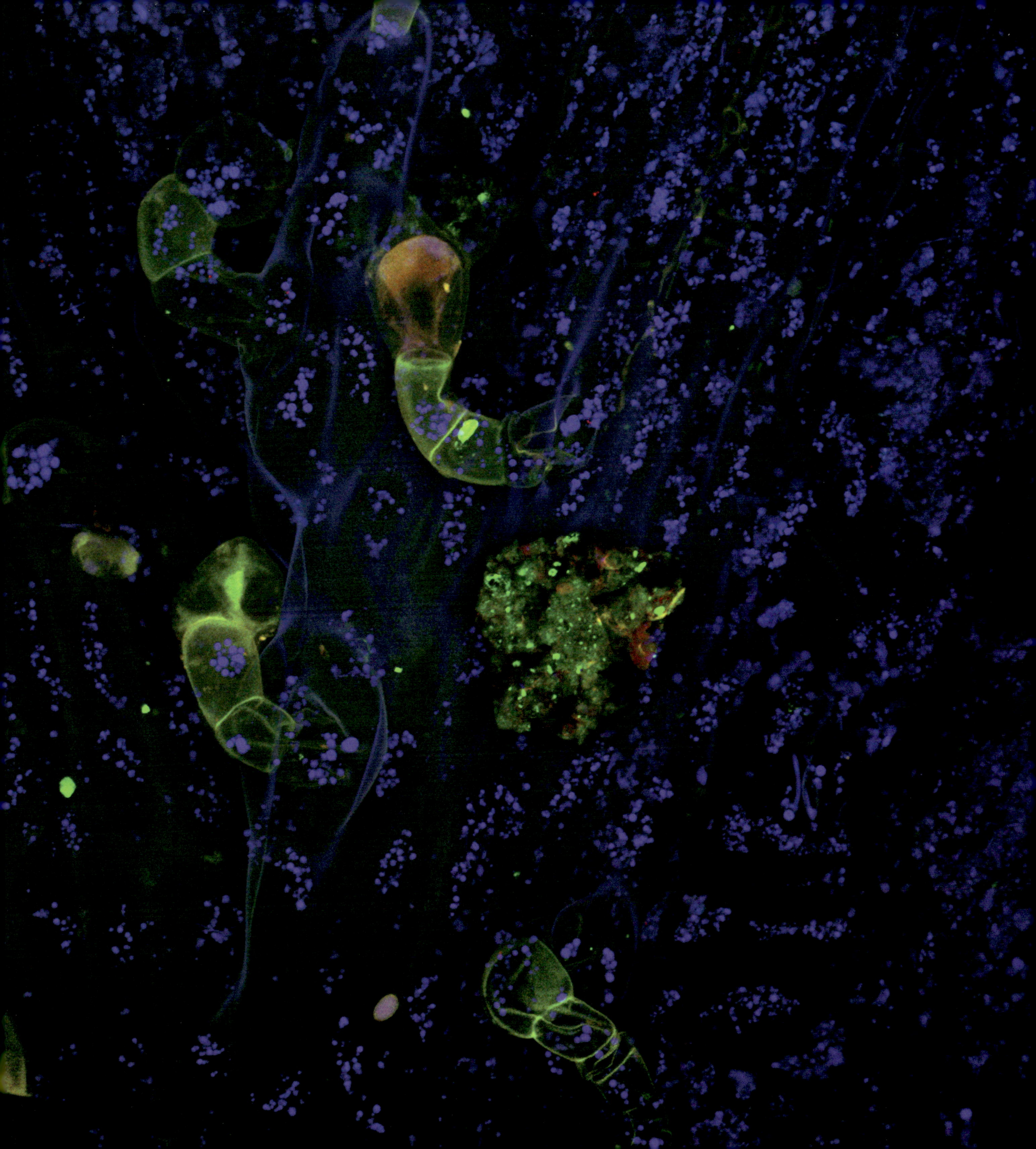

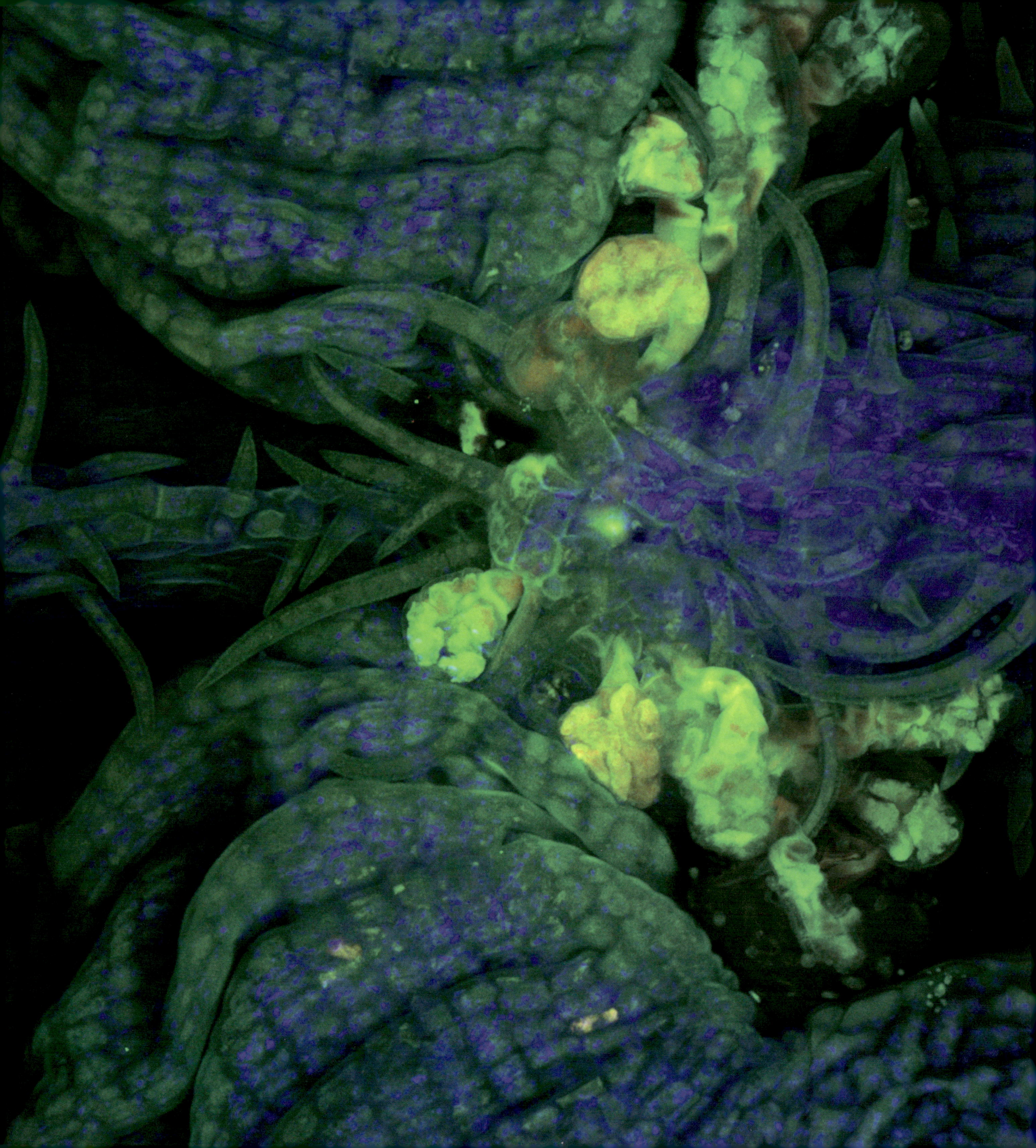

Fabaceae
Anadenanthera colubrina

Angico, cebil, vilca

The coin-like black seeds of this tropical and subtropical tree contain the powerful psychoactive alkaloid bufotenine, and have been used ceremonially by different Amerindian groups for thousands of years. It is closely related as a species to Anadenanthera peregrina, whose utilisation in shamanic snuff powders by the Taínos was documented in the 15th century in the Greater Antilles by Columbus.

Constantino Manuel Torres and David B. Repke, authors of the most comprehensive study of this plant, *Anadenanthera: Visionary Plant of Ancient South America*, maintain that "the genus *Anadenanthera* was, together with tobacco, one of the most widely used shamanic inebriants. It is primarily South American in distribution and includes two species with two varieties each. The earliest evidence for the use of psychoactive plants in South America is provided by remains of seeds and pods recovered from archaeological sites four millennia old. Seeds are roasted, pulverised, and inhaled through the nose, or smoked in pipes or as cigars".

They also point out that "the earliest descriptions of the use of visionary plants in the Americas refer to smoking of tobacco and inhalation of powdered seeds of *Anadenanthera peregrina* by the Taínos of the Greater Antilles [...]".

The first description of snuffing practices in the Americas was written by Christopher Columbus from observations made during his second voyage (1493-1496). During his brief period of residence on the island of Hispaniola, Columbus observed that the natives

right: *Anadenanthera colubrina* (cebil); seeds
opposite: *Anadenanthera colubrina* (cebil); confocal image

50 µm

engaged in a religious ceremony in which the snuffing of a psychoactive powder was an integral part.

Hundreds of thousands of examples of rock art possibly produced by the Carijona people have been identified at the archaeological site of Chiribiquete (located in the departments of Caquetá and Guaviare in Colombia). The oldest drawings could be up to 20,000 years old. The paintings in the rock shelters include depictions of sacred plants, among them the psychoactive acacia *Anadenanthera peregrina*, known as cohoba and yopo. Colombian archaeologist and anthropologist Carlo Castaño-Uribe, author of the indispensable study *Chiribiquete: la maloka cósmica de los hombres jaguar*, writes that: "in the sacred iconography, the yopo seed is represented with a spiky stem and a bifurcation for the beginning of germination. As the seedling grows, it forms a central branch (tridigit). In many representations the image is synthesised with a fully horizontal germination and ascending leafy ramifications, which are associated with dances with the Center of the World Pole, a key aspect of the dance rituals observed in Chiribiquete".

The images from the confocal microscope that we have included in this book are of *Anadenanthera colubrina*, which is from South America. The incontrovertible archaeological evidence in the form of actual seeds that Torres and Repke mention comes from sites in northern Chile and Argentina, as well as Bolivia, near Lake Titicaca. There are also extraordinarily artistic snuffing trays and other paraphernalia associated with the ingestion of the toasted and crushed seeds that contain high amounts of bufotenine.

I met Manolo Torres in 1983 when we both had Fulbright grants to work on projects in Chile. I had the rare privilege of seeing Manolo as he worked through his hypotheses and fascinating questions, still unresolved at that time, when he invited me to visit him in San Pedro de Atacama, one of the driest and most beautiful places on earth. Ancient mummies, snuff trays and, at night, more stars than I had ever seen. The Milky Way is a white river there!

The very closely-related species *Anadenanthera peregrina*, called cohoba by the Taínos in the Caribbean, was documented by a friar, Ramón Pané, who was commissioned by Columbus to study the ceremonies and antiquities of the Indigenous people who inhabited the islands. Pané, beginning in 1494, worked a full four years on his ethnographic research, which included specific references to this all-important psychoactive powder made from the seeds of *Anandenanthera peregrina*. The Inquisition incited the violent banning of this sacred plant along with the rituals associated with it that were considered a threatening source of Indigenous social coherence and unwanted competition with Christianity. This tragedy also marks the beginning of Europe's ecological devastation of the Americas. And, of course, the regional human toll with regard to the subsequent extermination of the Amerindian population of the Greater Antilles could not be greater.

Palaeoethnobotanic evidence discovered by a team of researchers led by Matthew E. Biwer during excavations at a site in Quilcapampa strongly suggests that the Wari culture during the Middle Horizon (AD 600-1000) produced a psychoactive fermented drink by combining *Schinus molle* drupes and *Anadenanthera colubrina* (vilca) seeds. According to this article published in the journal *Antiquity*: "vilca-infused molle chicha enabled a more inclusive psychotropic experience in Wari society. For perhaps the first time in the Andes, the consumption of vilca therefore moved beyond those spiritual leaders who communed with the supernatural realm". The public, ritualised partaking of this brew is an example of the ancient use of hallucinogens in Peru to coordinate collective action and create social cohesion.

In the extraordinarily insightful and comprehensive study "Contemporary Uses of Vilca (*Anadenanthera colubrina* var *cebil*): A Major Ritual Plant in the Andes", Verónica S. Lema, an anthropologist from the National University of Córdoba in Argentina, highlights "the enduring ritual value" of *Anadenanthera colubrina* from a

"The wide spatial and temporal distribution of the evidence for Anadenanthera ritual and visionary use attest to its importance in the construction and subsequent maintenance and modification of pre-Columbian and postcontact Indigenous ideologies."

~ *Constantino Manuel Torres*

"New chemical and microbotanical analyses conducted by a team of researchers led by anthropologists John W. Rick and Verónica S. Lema confirm for the first time the ritual use of Anadenanthera colubrina (cebil) and Nicotiana at the ceremonial centre of Chavín de Huantar in Peru as early as 1000 BCE. The scientists affirm that "use of psychoactive plants was intimately associated with institutionalised ritual, and not limited to the individualised contexts of ecstatic shamanism."

Anadenanthera peregrina (yopo); plant

above: *Anadenanthera colubrina* (cebil); Argentina
opposite: *Anadenanthera colubrina* (cebil); confocal image
50 μm

overleaf: *Anadenanthera colubrina* (cebil); confocal image
25 μm

pre-Hispanic past to the contemporary south Andean world. Vilca or cebil, says Lema, is used for magical-religious, medical, and veterinary purposes, as well as for construction, fuel, fodder, dyeing, and artefact-making. In keeping with Andean conceptions of illness, Lema describes how Vilca seeds are believed to "act as protective amulets, playing a dual role: shielding the body to prevent the displacement of its spirit and embodying a continuous, rotation movement, [which] compels incoming negativity to reverse its trajectory and to return to its point of origin". Lema provides detailed information about the multiple ways that *Anadenanthera colubrina* is used not only for protection and to bring good luck, but also as a purge, cleanser, medicine, and ritual drink, as well as an ingredient in ritual bundles for ceremonial altars (*mesas*). Lema undertook this fieldwork between 2017-2019 by conducting interviews with people selling medicinal products at numerous traditional markets in Peru, Bolivia, and northwest Argentina.

A group of scientists, primarily from Brazil's Universidade Federal de Mato Grosso, and led by Merline Delices, published an overview of *Anadenanthera colubrina* that demonstrates how recent pharmacological studies corroborate popular therapeutic uses of extracts of this plant to heal wounds and as an anti-inflammatory, antioxidant, antidiarrheal, antifungal, and antitumoral. The scientists warn that unregulated use of bark and seeds from this plant for medicines and for recreational psychedelic experiences may result in its extinction.

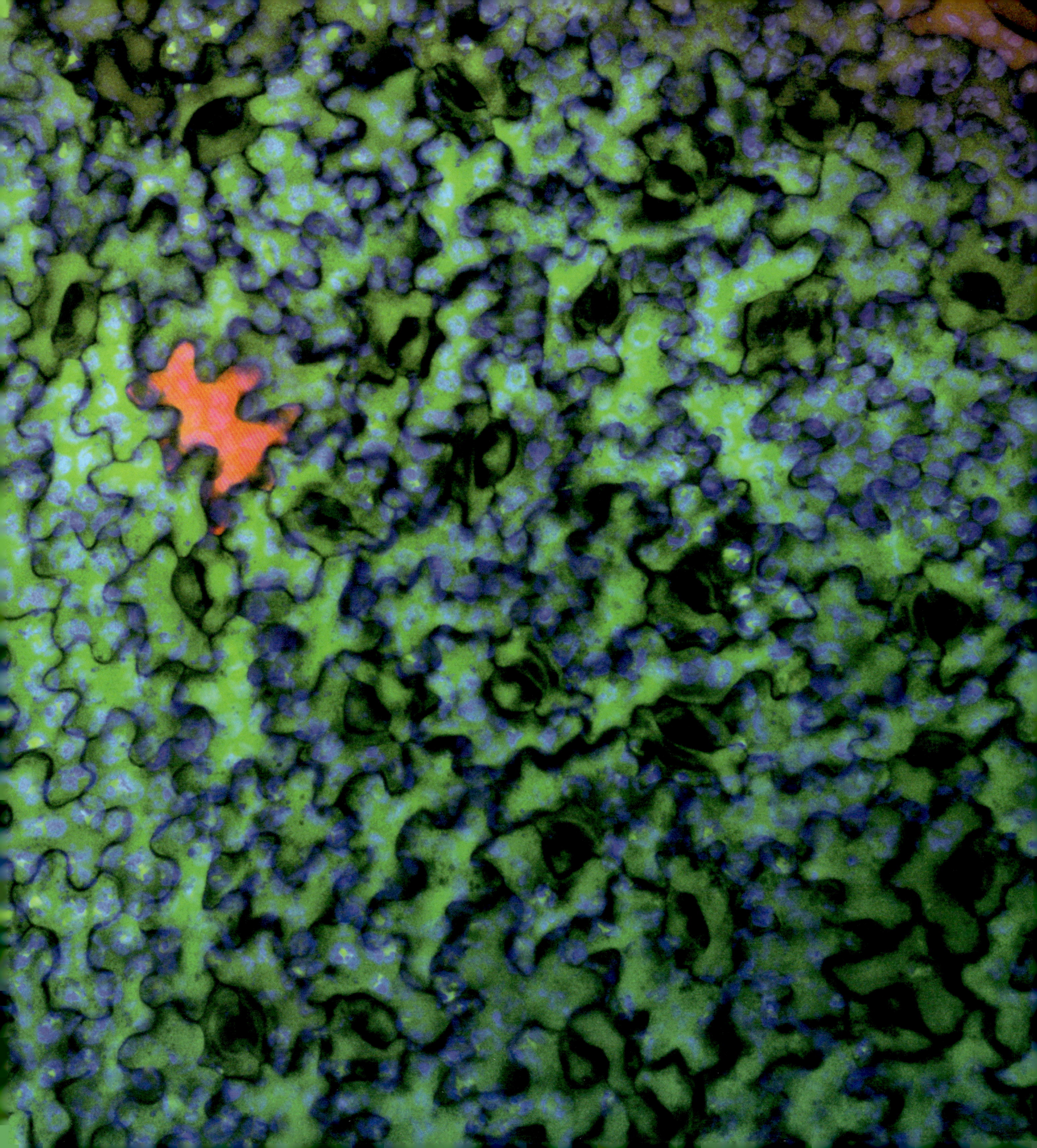

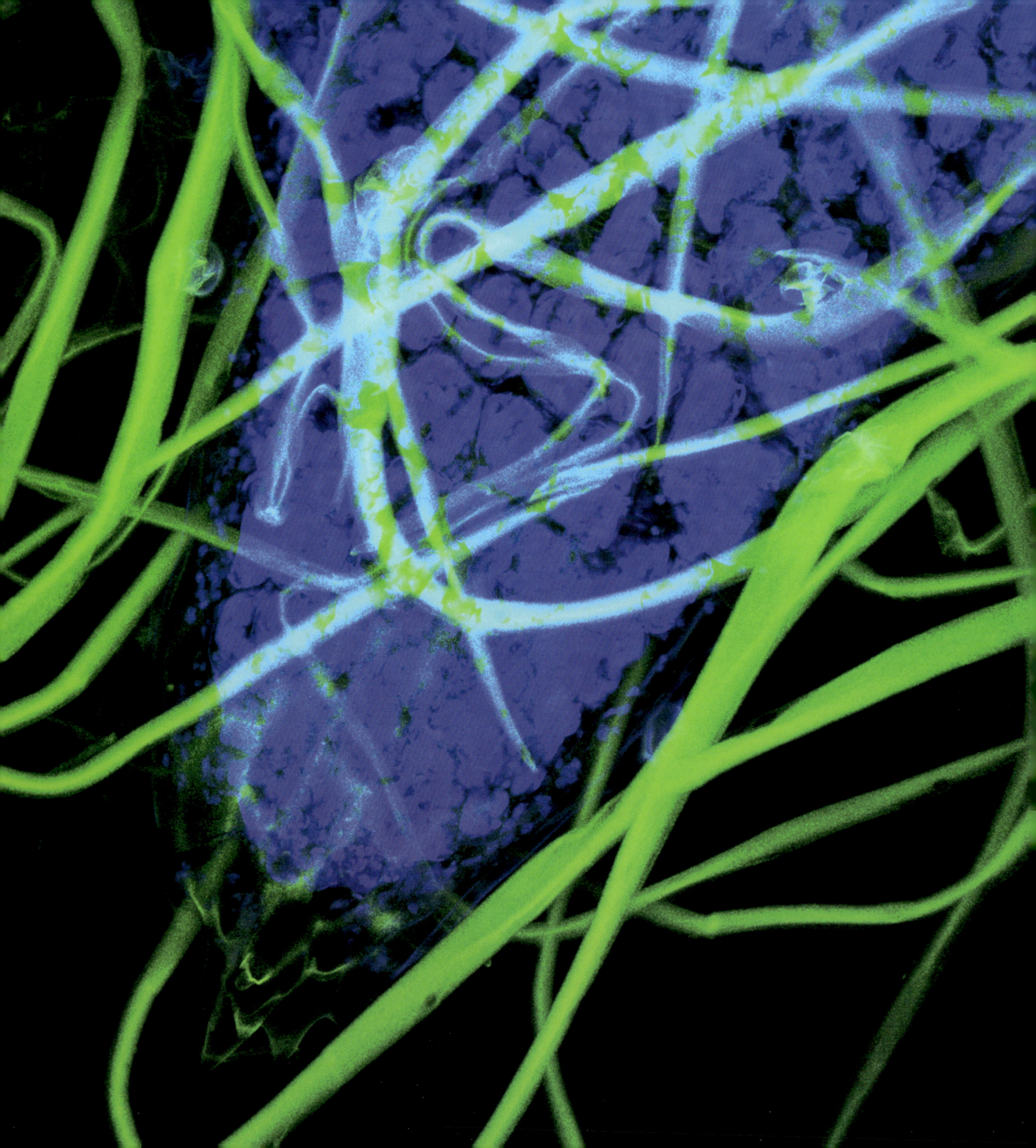

Asteraceae
Artemisia ludoviciana subsp. *mexicana*
Estafiate, Mexican white sagebrush, western mugwort

This perennial plant with small, whitish-grey silky leaves grows in dry, rocky subsoil. It was used as a ritual incense by the Plains Indians. Contemporary scientific research is demonstrating that its essential oils have a wide range of important medicinal properties.

A*rtemisia ludoviciana* ssp. *mexicana* (known commonly as western mugwort) is found throughout the Southwestern US in addition to both the dry and warm zones of Mexico.

In the monumental *Encyclopedia of Psychoactive Plants: Ethnopharmacology and Its Applications*, Christian Rätsch describes the North American prairie sagebrush *Artemisia mexicana* as the "most important ritual incense of the Plains Indians", for whom the rising fragrant smoke "links together Maká, the Mother Earth, with Wakan Tanka, the Great Spirit, who is active in all creatures". The plant was also used for this ceremonial purpose by the Aztecs in the pre-Columbian era, and is mentioned in the Florentine Codex as being associated with Uixtociuatl, the Aztec goddess of salt and salt makers. In ritual dances, the staff of Uixtociuatl is adorned with wormwood leaves, whilst the participants, who are connected by a flower rope, also wear wormwood flowers in their hair. This plant is also held sacred to Tláloc, the god of rain.

In *Pharmacotheon*, Jonathan Ott summarises research demonstrating how different species of *Artemisia* were used as traditional analgesics and stimulants by the Zuni, the Cheyenne, and the Potowatomi. Ott also says that "the ancient Aztecs used *Artemisia mexicana* as an inebriant, under the name *itzauhyatl*", and cites sources linking this plant to

Artemisia ludoviciana ssp. *Mexicana* (western mugwort); confocal image

50 μm

the sacraments *peyótl* and *ololiuhqui*. Estafiate, the Spanish name for *Artemisia ludoviciana* subsp. *mexicana*, is an ethnomedicine in current use by urban Mexicans as well as the Tarahumara. For their part, the authors of *Plants of the Gods* document the presence of a bundle of sagebrush (*Artemisia*) for smudging purposes among the roadman's essential ritual implements for conducting peyote ceremonies in the Native American Church.

Scientific studies directed by Gerardo D. Anaya-Eugenio indicate that "*Artemisia ludoviciana* preparations showed hypoglycemic and antihyperglycemic effects, which could explain its effectiveness for treating diabetes in contemporary Mexico". Subsequent research led by Anaya-Eugenio with regard to this plant's widespread use as a popular remedy in Mexico confirm that "essential oils from a wide range of *Artemisia* species have been largely employed for their antiinfective, analgesic, antipaludic, anticancer and anti-inflammatory alleged properties". Based on the experiments conducted, the article concludes: "the neurogenic and peripheral antinociceptive effects of the essential oil of the plant were demonstrated; since these effects were partially blocked by naloxone, an opioid mechanism action was proposed". A team of researchers from Mexico led by Juan Francisco Palacios-Espinosa conducted a 2021 study of *Artemisia ludoviciana* subsp. *mexicana* that "validates traditional consumption methods" of the plant due to its "gastroprotective and anti-inflammatory activities". The scientists call their work on *Artemisia ludoviciana* as a source for antibiotics against *Helicobacter pylori* a "remarkable contribution to the ethnopharmacological knowledge of this species". José Luis Gálvez Romero led a group of scientists that published a study in 2022 confirming the antimycobacterial activity of *Artemisia ludoviciana* and suggesting that ethanol extracts of the plant "could potentially be used to supplement the treatment of tuberculosis".

Artemisia ludoviciana ssp. *Mexicana* (western mugwort); Montana

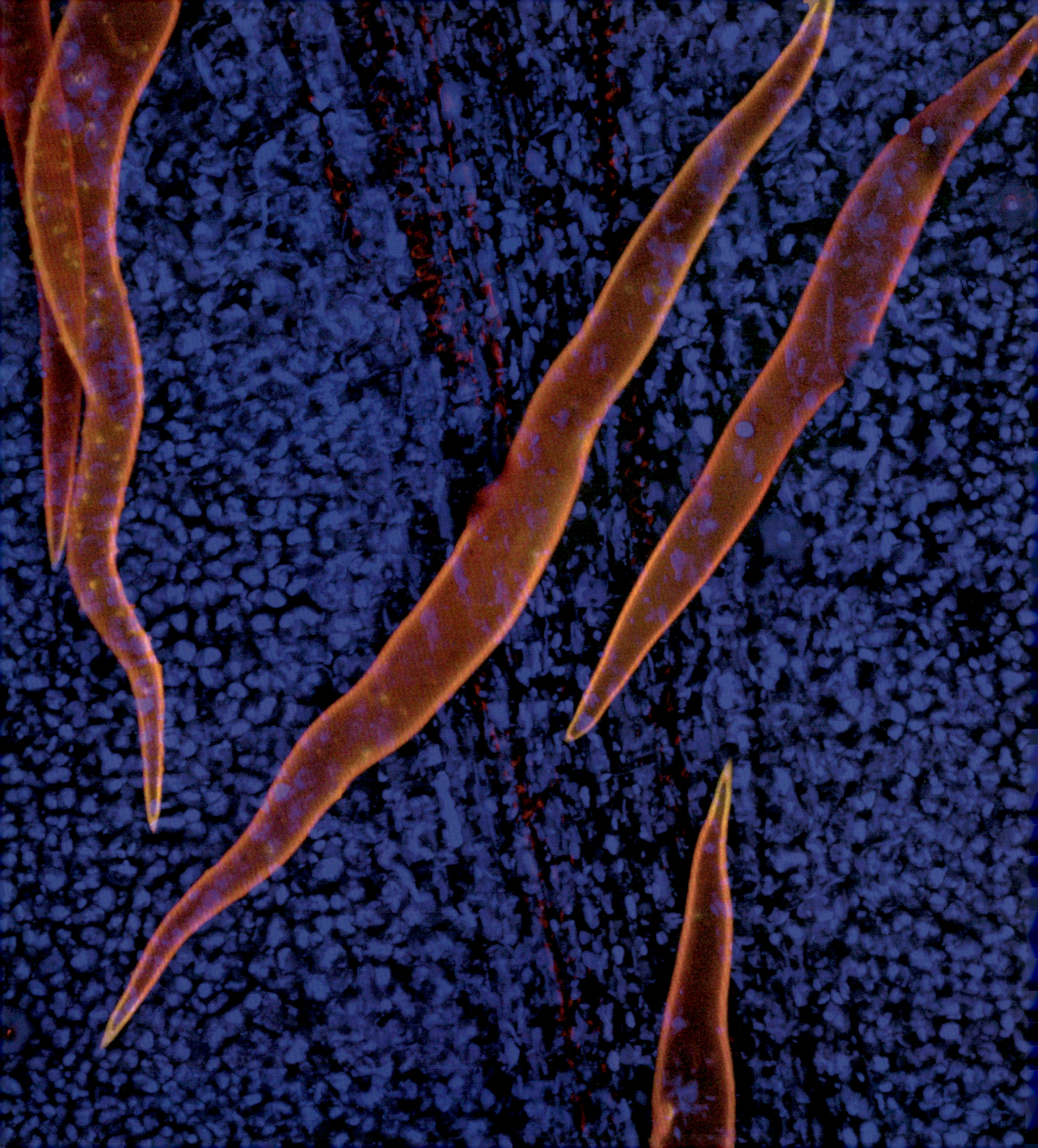

Ayahuasca / yagé

the combination of:

Malpighiaceae
Banisteriopsis spp.

Ayahuasca, hoasca, jagube, yagé, miiyabu, red ayahuasca

This giant tropical liana that seeks the support of surrounding trees to climb is the basis of a sacred beverage called ayahuasca, or yagé, that is used by highly-prepared Amazonian visionary doctors to engage with spirits from suprasensible worlds, to converse with ancestors, and to diagnose illness in their Indigenous communities. There is a great deal of current scientific research seeking ways to distinguish between the numerous ethno-varieties of Banisteriopsis caapi.

Malpighiaceae
Diplopterys spp.

Chagropanga, chaliponga, huambisa

Two species of the scandent vine Diplopterys, a source of the psychoactive substance DMT, are a common admixture to a sacred drink that always includes Banisteriopsis caapi. Diplopterys cabrerana (chaliponga and oco-yagé) is used for the brew in Colombia. Diplopterys longialata (huambisa) is often utilised for this purpose in southern Ecuador and, increasingly, in Peru as a more ecologically adaptable substitute for Psychotria viridis.

Rubiaceae
Psychotria spp.

Amyruca, chacruna, kawa

Psychotria viridis (known as chacruna) is a member of the coffee family with similar ovate shiny leaves. It is the preferred DMT-source admixture plant in the ayahuasca sacramental drink in parts of Peru and throughout Brazil, where the Santo Daime church cultivates different varieties. Psychotria carthagenensis (amyruca) is sometimes added by traditional Lamista healers in Peru to their preparations of ayahuasca.

For the sake of clarity, we have decided to create a special grouping of sacred plants for *Microcosms*. *Banisteriopsis* spp., *Diplopterys* spp., and *Psychotria* spp. are the plants most widely used to create the sacred drink known either as ayahuasca or yagé, depending on the brew's geographical origin. Our organising principle here for this book comprised of many different species of sacred plants, however, is "B" for *Banisteriopsis*. To be clear, the two words "ayahuasca" and "yagé" are used to name a drink that is composed of more than one plant. But both words also designate the single vine *Banisteriopsis caapi* on its own. Neil Logan, writing in "The Yagé Complex", explains the importance of highlighting *Banisteriopsis caapi* as the common underlying element that joins all the multiple variations possible in the preparation of the sacred beverage: "ultimately, the use of *Banisteriopsis caapi* combined with more than one hundred potential admixture plants, became common across the eastern Andes from Bolivia, north to Colombia, and Venezuela, following the Amazon and its tributaries eastward across much of north and central Brazil. "Caapi" or "cabi" are two of the more common names for referring to related vines across most of northern South America. *Banisteriopsis caapi* is considered by many groups of these regions to be a kind of driver of ecological ingenuity. It is the fundamental master medicinal plant teacher around which all other plants revolve".

"Sometimes they mix tara yagé, waˊi yagé and pehí so that the result is very concentrated. When you drink it, the drunkenness hits you before you finish the gourd. You feel burns all over your body like you're being hit with burning logs. Then the body catches on fire and is reduced to ashes. When the flesh is destroyed, only then does the soul emerge and begin to see. At that moment the most fantastic visions begin."

~ Fernando Payaguaje, The Yagé Drinker

Constantino Manuel Torres summarises the synergy between these plants in his brilliant study "From Beer to Tobacco: A Probable Prehistory of Ayahuasca and Yagé": "the *Banisteriopsis* vine contains several -carboline alkaloids – harmine, harmaline, and tetrahydroharmine – which are potent inhibitors of the enzyme monoamine oxidase (MAO). Frequently, ayahuasca and yagé are combined with the leaves of *Psychotria viridis* (Chacruna) or *Diplopterys cabrerana* (Chaliponga, Chagropanga, Oco-Yagé). The leaves of these two species contain N,N-dimethyltryptamine (DMT), which is not orally active. However, its combination with the MAO-inhibiting harmala alkaloids allow for its activity". Therefore, although these species could very well be considered separately, they are more conveniently regarded together as a sacred synergetic mixture.

We are pleased to offer some additional botanical information that is rarely given the attention it deserves. In some brief comments on his photos of *Diplopterys longialata* (huambisa), Alan Rockefeller maintains how important it is to keep in mind the overlooked or even occluded presence of this particular species as perhaps the most common plant additive in the ayahuasca brew always prepared with *Banisteriopsis caapi*. Often, he affirms, *Diplopterys longialata* is misidentified as *Diplopterys cabrerana*. Indeed, botanically, the two species of *Diplopterys* can easily be confused if they are not flowering.

page 54: *Banisteriopsis caapi* (ayahuasca / yagé); confocal image
opposite: *Psychotria viridis* (chacruna); confocal image

50 μm

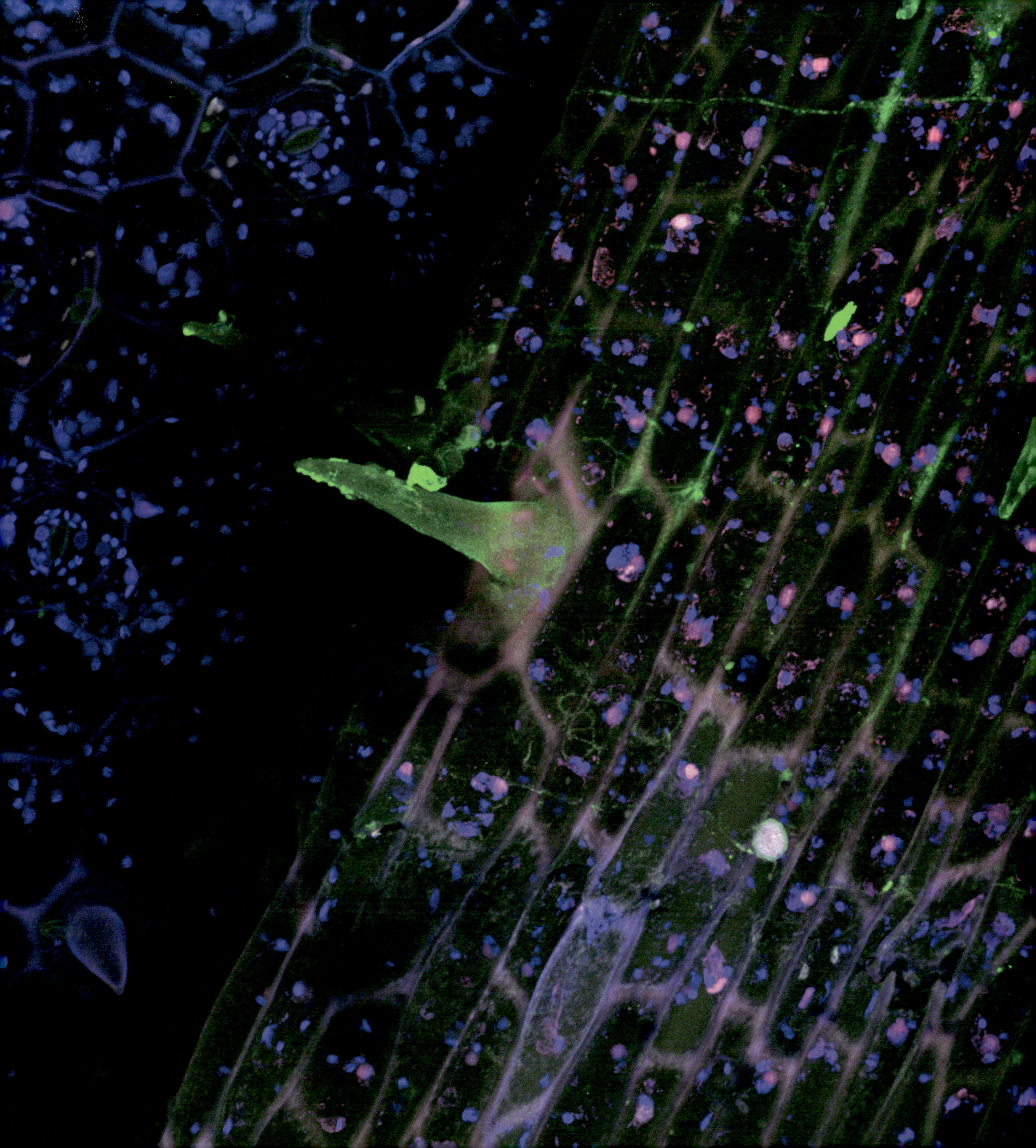

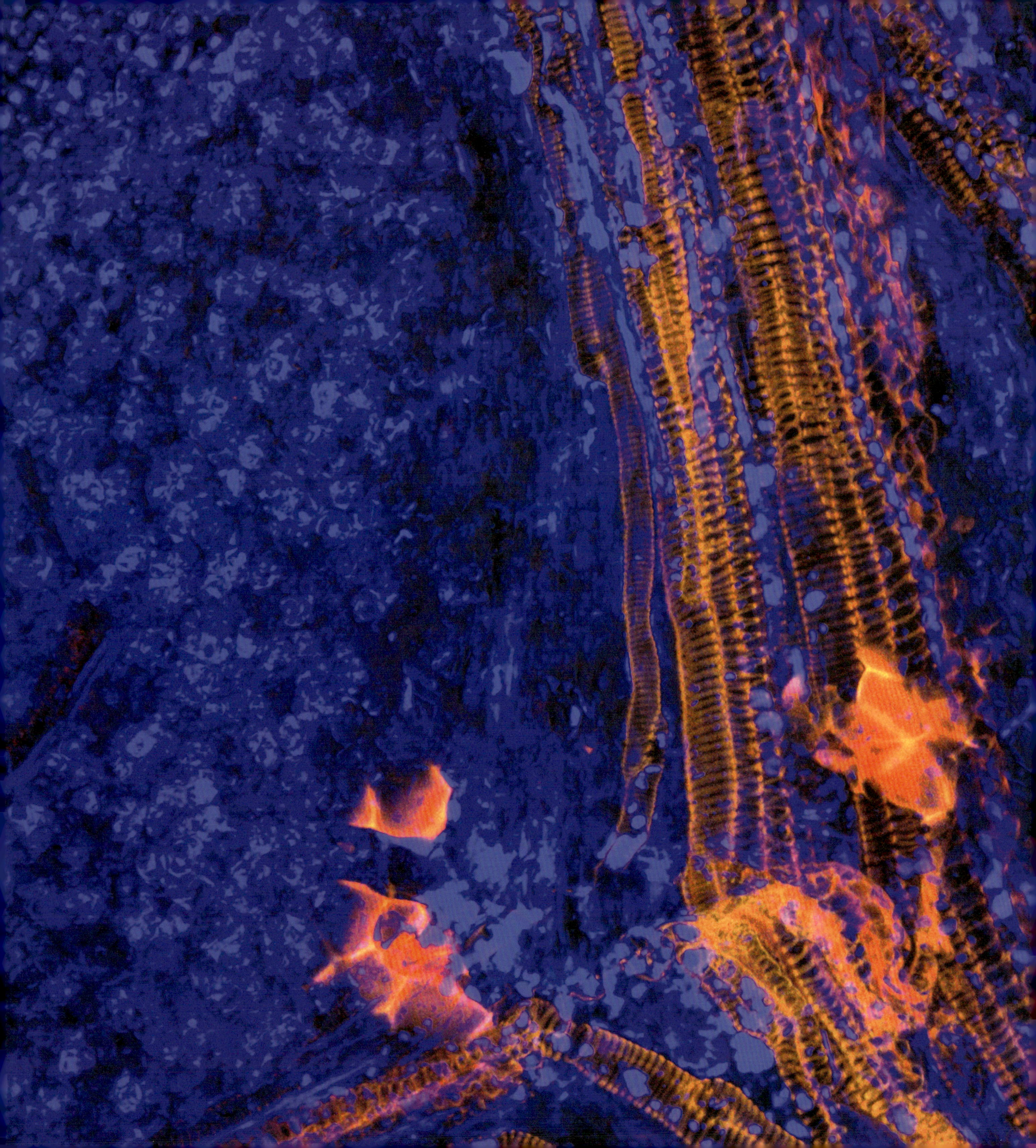

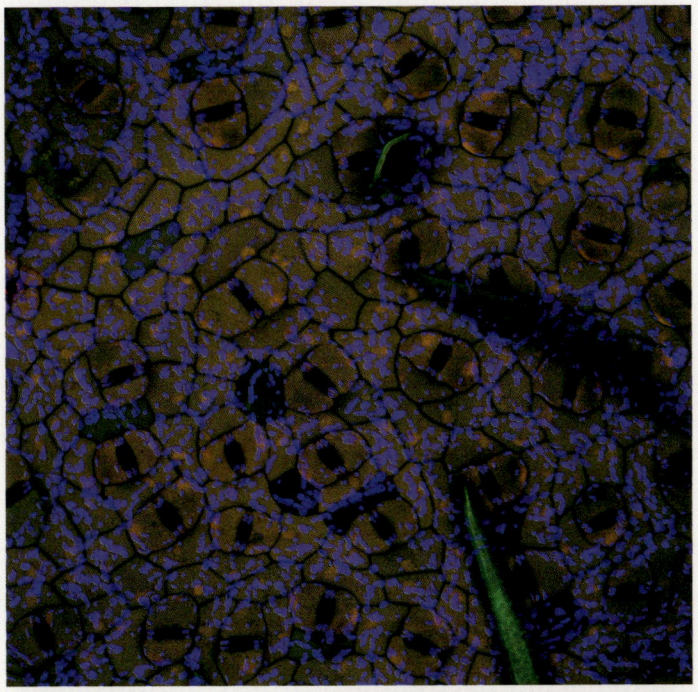

top: *Diplopterys cabrerana* (chaliponga); confocal image

50 μm

above: *Diplopterys cabrerana* (chaliponga); plant, Costa Rica
opposite: *Banisteriopsis caapi* (ayahuasca / yagé); confocal image

50 μm

However, generally speaking, *Diplopterys cabrerana* is almost always used to prepare yagé in Colombia. In northern Ecuador along the Colombian border, *Diplopterys cabrerana* and *Diplopterys longialata* are used with *Banisteriopsis caapi* interchangeably, and sometimes together. *Diplopterys longialata*, known by its common name huambisa, is used in ayahuasca preparations in southern Ecuador. Additionally, over the last fifty years, it has been introduced into Peru, becoming increasingly popular as a substitute for *Psychotria viridis* due to its similar entheogenic strength/quality, and also because it is more resilient in terms of climate fluctuations such as cold, drought, and flooding.

In Brazil, ayahuasca is also known as daime, a sacrament used by members of the Santo Daime church, which has legal status and exists throughout the country. As always, the preparation includes the obligatory *Banisteriopsis caapi* (known also as jagube) and, in Brazil, the plant admixture *Psychotria viridis* (called "a Rainha" – "the Queen" by the *daimistas*). Life, of course, is complicated on account of biodiversity. So, a curious reader with some Brazilian Portuguese can read the doctoral dissertation by Ricardo Monteles about the different varieties of sacred plants used in the Santo Daime ceremonies. Also in Brazil, Regina Célia de Oliveira is undertaking serious scientific studies with other academic researchers on the numerous *Banisteriopsis caapi* ethno-varieties. Unfortunately, the names of these ethno-varieties in Brazil do not coincide with the plethora of Indigenous names for varieties of *Banisteriopsis caapi* (including wai yagé, tara yagé and tzinca) in combination with *Diplopterys cabrerana* (oco yagé) in the Northwest Amazon, which is the probable geographic origin of the synergistic plant knowledge that, over time, evolved in the following way: from *Banisteriopsis caapi* used on its own, to chewed *Banisteriopsis caapi* raw stems combined with the ground seeds of *Anadenanthera peregrina* (a source of bufotenine, 5-OH-DMT), to *Banisteriopsis caapi* stems boiled with the leaves of *Diplopterys cabrerana* (which may have begun as recently as less than 200 years ago, according to Torres). We are pleased to offer in *Microcosms* confocal images of some of these Amazonian legacy vines. Not all *Banisteriopsis caapi* is the same. Hardly! The great Richard Evans Schultes may have had difficulty distinguishing between these varieties of the sacred Amazonian vine, but this is not true for the Siekopai, Siona, and Cofán with their sophisticated ethno-taxonomy. Jonathon Miller Weisberger has studied this phenomenon in *Rainforest Medicine: Preserving Indigenous Science and Biodiversity in the Upper Amazon.*

Luis Eduardo Luna and I met at Palenque in 1996 for a gathering sponsored by the Botanical Preservation Corps and began the structural planning for what would become the nearly 500-page volume *Ayahuasca Reader: Encounters with the Amazon's Sacred Vine.* From the onset, it was the highest priority for us as co-editors to emphasise what might be called an Indigenous research paradigm. *Ayahuasca Reader*, like *Microcosms*, is a tribute to the Amerindian receivers, keepers, and perpetuators of particular vegetal lives that are gifts from the gods. For this reason, the first of five different sections in the anthology is called "Ayahuasca Myths and Testimonies", and collects plant narratives related to *Banisteriopsis caapi*, *Diplopterys* spp., and *Psychotria* spp. Sometimes, as in the case of Gerardo Reichel-Dolmatoff, the stories reach us through old-school, now questionable, anthropological methodologies using anonymous informants to create paraphrased recreations. In other instances, the ethnographers provide more information, and, rightly so, furnish the names of Indigenous guardians of shamanic tales such as Ricardo Yaiguaje (Siona), Milton Maia and Maria Domingo (Cashinahua/Huni Kuin), Mengatue Baihua and Huepe Orengo Coba (Huaorani), Alberto Prohaño (Yagua), Hilario Peña (Inga), and, finally, Fernando Payaguaje (Secoya/Siekopai), the extraordinary *bebedor de yagé* (yagé drinker), whose extensive and invaluable first-person testimony was preserved in Payaguaje's first language Pai-Coca by the very elderly healer's grandchildren, then translated into Spanish. One hopes that these voices (recorded as interviews, transcribed, edited, translated, and even translated yet again into a third language) are ethically and equitably collected. One deeply appreciates these words, even as one recognises that there is always a complex process of mediation occurring that involves close family members fighting oblivion in the inexorable flow of time or a foreign anthropologist, perhaps a graduate student hoping to finish a dissertation or someone such as Bruce Albert, who collaborated with healer and activist Davi Kopenawa over decades to create the remarkable book *The Falling Sky: Words of a Yanomami Shaman.*

Despite these filters, nonetheless, the plants are able to make themselves known. An awareness of inevitable mediating processes also makes one personally cherish less mediated contact, in my case a long direct conversation with the highly-respected Onanya (Shipibo visionary doctor) Don Benito Arévalo in Pucallpa, Peru in June, 2000. It was a true privilege to talk with him about these healing plants and then watch or, rather, hear him labour through the entire night as he treated local patients (none of whom drank ayahuasca) for a wide variety of maladies. In this Shipibo context, it was the doctor, not the patient, who drank ayahuasca in order to diagnose and cure difficult and persistent illnesses.

Pedro Favaron's enormously insightful books *Las visiones y los mundos: sendas visionarias de la Amazonía Occidental* and *La senda del corazón: sabiduría de los pueblos indígenas de Norteamérica* are journeys into traditional Indigenous knowledge. Favaron (a Peruvian-Argentine of Italian descent from Lima and the Shipibo-Konibo Native Community of Santa Clara) is married to the accomplished Shipiba artist Chonon Bensho from Santa Clara de Yarinacocha, Peru, and, through her family, now also his, he is able to describe lineages of legendary healers (such as his wife's grandfather Ranin Bima) and their relationship to plant medicine, ancestral narratives, and songs by engaging in dialogue with other members of his family (especially his father-in-law Menin Bari and his uncle Kene Jisman) over the long periods of time that constitute lifetimes of shared responsibilities and accumulated knowledge. The visionary doctors of the Shipibo nation, according to Favaron, undergo arduous initiations that enable them to establish relations with the Ibo, or Dueños (in the double sense in English of both Owners and Masters) of the medicinal plants called *rao* in the Shipibo language. It is thanks to these plant-alliances created through ritual dieting that the traditional physician is able to use songs to cure in keeping with the healing powers of particular species. In "Netabaon Joi: the Shipibo-Konibo

opposite top left: *Banisteriopsis caapi* (ayahuasca / yagé); Ecuador
opposite bottom left: *Banisteriopsis caapi* (ayahuasca / yagé); Ecuador
opposite right: "Transformation of Taita Rufino", oil on wood painting by Colombian artist Jeisson Castillo

Ayahuasca / yagé

Cosmic Semiotics", Favaron concludes that "the diverse beings of the cosmos are all interwoven in a single communicating loom".

In general, Favaron has a very negative view of the explosive increase in the non-Indigenous globalised use of ayahuasca in recent decades. As he puts it, "the visionary medicine of the Western Amazon has become the new spiritual territory that the modern way of thinking wants to profane and commercialise". He laments the confusion and the lack of respect that he perceives in relation to ayahuasca, and recognises that "some Indigenous persons, with little preparation, call themselves maestros just to do business". "The Shipibo doctors in the olden days", continues Favaron, "did not have the custom of giving ayahuasca to their patients, but, rather, drank to connect themselves to spiritual worlds and cure the sick by singing songs and using other medicinal plants".

Alex K. Gearin, author of *Global Ayahuasca: Wondrous Visions and Modern Worlds*, analyses the burgeoning use of ayahuasca in several contexts, including non-Indigenous foreigners arriving en masse to the Peruvian Amazon for spiritual retreats with Shipibo healers, who are contracted to provide these services in businesses that are owned by foreign nationals. The guests, called pasajeros, or passengers, by their hosts, writes Guerin, "came to heal themselves, learn about their own spiritual interior, and transcend 'modern' problems with shamans seen to be relatively uncorrupted by the ills of civilisation". These centres, however, are based on what Gearin calls "a double dislocation": "Indigenous healers are dislocated from the place, context, and moral order of their existing local shamanic practices, and ayahuasca tourists are dislocated from their homelands and ordinary cultural realities when embarking on pilgrimages to the Amazon rainforest". Even so, ultimately, affirms Gearin, "ayahuasca has attracted people from distant corners of the planet precisely because of its adaptive ontological capacities". His study also documents ayahuasca use in Australia (where Australian facilitators guide ayahuasca drinkers who "aim to heal distress and sickness by imbibing a natural antidote sometimes said to heal the trauma of society itself") and, yes, believe it or not, in China (where users tend to be young, wealthy Chinese entrepreneurs and corporate managers "searching for holistic wellness, self-cultivation, and a competitive edge in capitalist environments"). During his research and interviews in China, what struck Gearin, who teaches

top: *Psychotria viridis* (chacruna); Nicaragua
centre: *Psychotria carthagenensis* (amyruca); Nicaragua
bottom: *Psychotria viridis* (chacruna); seeds
opposite: *Psychotria viridis* (chacruna); confocal image

50 μm

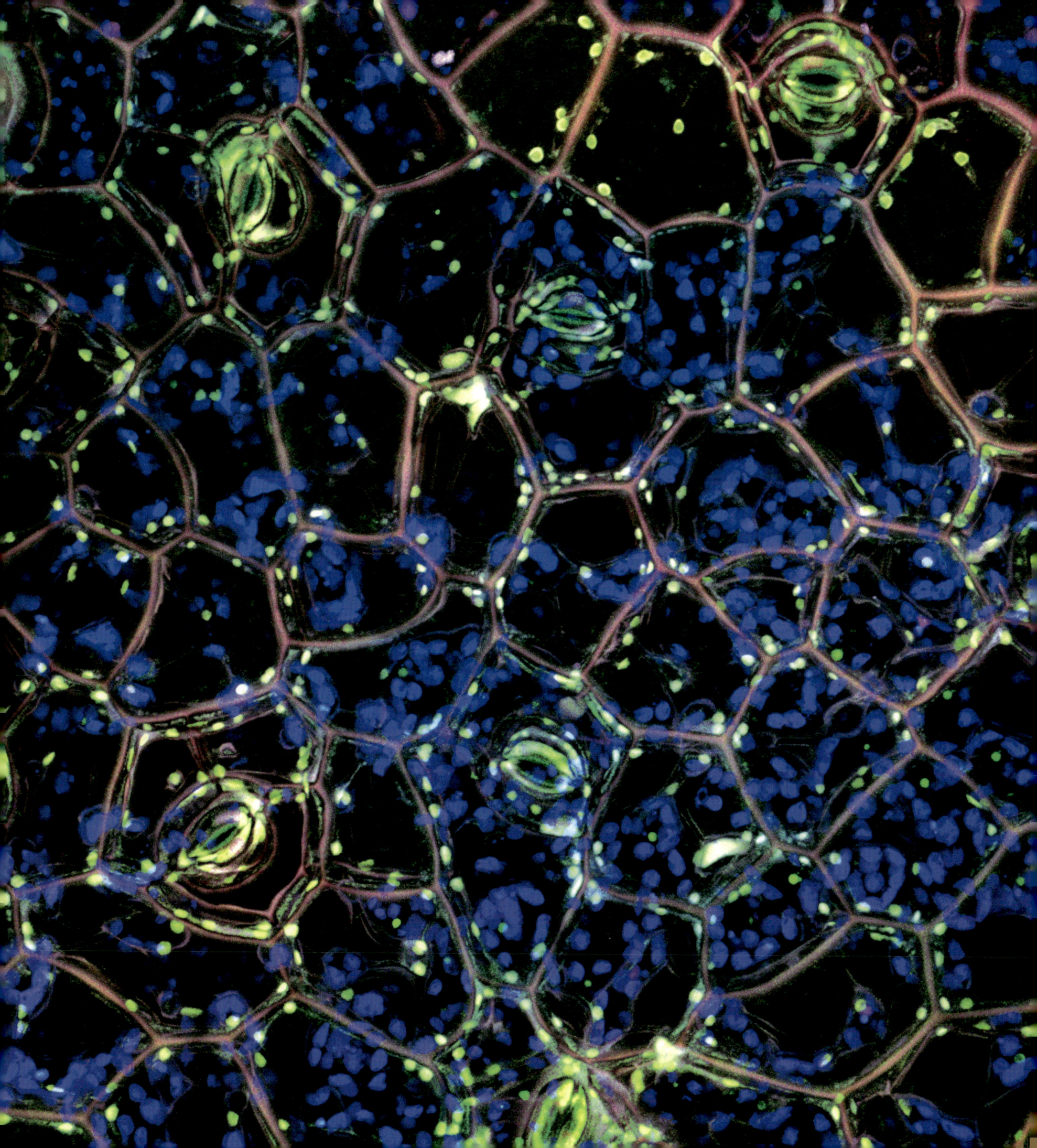

"On the other hand, there were other customs that got lost: for example, the yagé ceremony. When she learned of it, the missionary went around repeating: "It's bad to drink yagé, it's harmful." At that point, some people became evangelicals. I couldn't build another yagé hut on my own; so we let it go, despite the fact that I can still heal people."

~ *Fernando Payaguaje, The Yagé Drinker*

in the Medical Ethics and Humanities Unit at the University of Hong Kong, was the "utilitarian ethos of ayahuasca", and "the sanitisation of ayahuasca into a secular framework". In China, writes Gearin, psychoactive plants have "become a visionary technology employed to advance business life". Some insider will no doubt soon publish a book on the prevalence of ayahuasca use in the Hollywood film industry, the creation of Artificial Intelligence, and venture capitalism in Silicon Valley. Welcome to the contemporary world of global ayahuasca!

The astonishing growth in worldwide interest in ayahuasca has become a recurring theme in the mainstream media and in prominent publications such as David Wallace-Wells' bestselling *The Uninhabitable Earth: Life After Warming* (2019), in which the author describes a burgeoning Wellness Movement, saying, "what has been called the "new New Age" arises from a similar intuition – that meditation, ayahuasca trips, crystals and Burning Man and microdosed LSD are all pathways to a world beckoning as purer, cleaner, more sustaining, and perhaps above all else, more whole. This purity arena is likely to expand, perhaps dramatically, as the climate continues to careen toward visible degradation…".

Researchers such as Luis Eduardo Luna and Dennis J. McKenna, who have been writing for decades on this phenomenon that has been called an Archaic Revival, emphasise the transformative ecological perspectives that many people experience – and it's not always pretty! Luna describes how ayahuasca can "increase fully-sensed body-and-mind awareness of the current perils of environmental destruction, nuclear disaster, and social turmoil". McKenna proposes ayahuasca as a teacher, an "ambassador from the community of species", and, most importantly, "a catalytic influence in changing global environmental consciousness". In this regard, ayahuasca might propitiate a visceral, indelible, impassioned understanding of the term "biophilia", a love of life worth defending against its powerful enemies.

Additionally, as Dale Millard points out, the healing properties of harmine in ayahuasca are of utmost consideration. Millard's research overview demonstrates its "wide variety of therapeutic activity inducing antimicrobial, anti-diabetic, anticancer, antidepressant, antiparasitic, DNA-

below: *Diplopterys cabrerana* (chaliponga); plant, Costa Rica

above and centre: Miguel Payaguaje making cuttings of wai yagé to plant in a garden with pineapples, Sucumbíos Province, Ecuador, 2017

above: Fernando Payaguaje, Siekopai author of *El bebedor de yagé* (*The Yagé Drinker*)

binding, osteogenic, chondrogenic, neuroprotective and other effects. Harmine is by far the most abundant constituent of the medicine ayahuasca. Its presence in pharmacologically active amounts may therefore provide a rationale for its contribution in ayahuasca's wide application in traditional medicine and its general reputation for treating a broad range of diseases and ailments".

Psychotria viridis is the species of *Psychotria* that is the preferred ayahuasca admixture plant, though there is evidence that the closely-related species *Psychotria carthagenensis* also is used, especially by the formidable Lamista shamans in Peru, according to University of Cambridge medical anthropologist Françoise Barbira Freedman in her study "Shamanic Plants and Gender in the Healing Forest". Barbira Freedman affirms that "shamanic plant knowledge acquisition involves the understanding of the dynamic relations between the gendered species and the engineering of balance among them". She goes on to explain that there are androgynous trees as well as some plants that are not gendered: "for instance, the various plants that are labelled Ayahuasca (several varieties of *Banisteriopsis* and *Brugmansia*) are paired with plants that activate the visionary quality of the brews. These plants are generically called chacruna; the most commonly used species are two shrubs (*Psychotria viridis* and *Psychotria carthagenensis*) and a scandent vine (*Diplopterys cabrerana*)". It is interesting to note that, etymologically, the word chacruna is from the Quechua verb *chakruy*, which means to mix. In this important region of shamanic traditions, chacruna is not solely associated with *Psychotria viridis* (as it is elsewhere) but has, instead, a generic use and refers to a range of ayahuasca admixture plants. Despite certain controversies regarding the actual alkaloidal content of *Psychotria carthagenensis* in the context of phytochemical lab testing, the ritual Amerindian use of this species of *Psychotria* is well documented. For this reason, we include *Psychotria carthagenensis* (amyruca) among the sacred plants of *Microcosms*.

overleaf left: *Banisteriopsis muricata* (red ayahuasca); confocal image
overleaf right: *Diplopterys longialata* (huambisa); plant

50 μm

Ayahuasca / yagé

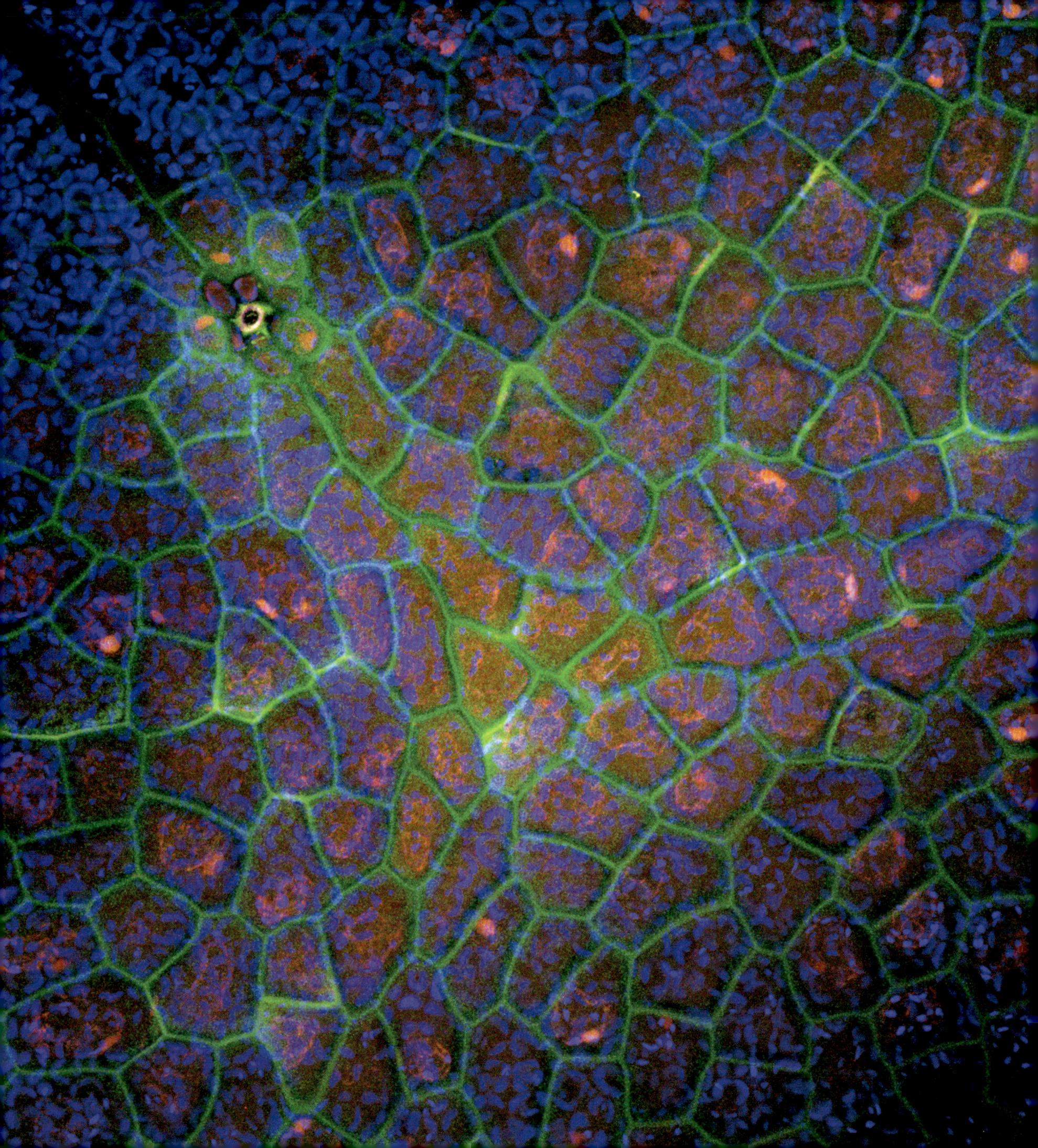

above: "Song of Ayahuasca", embroidered cloth, anonymous Shipiba artist, Peru
opposite: *Diplopterys cabrerana* (chaliponga); confocal image

50 μm

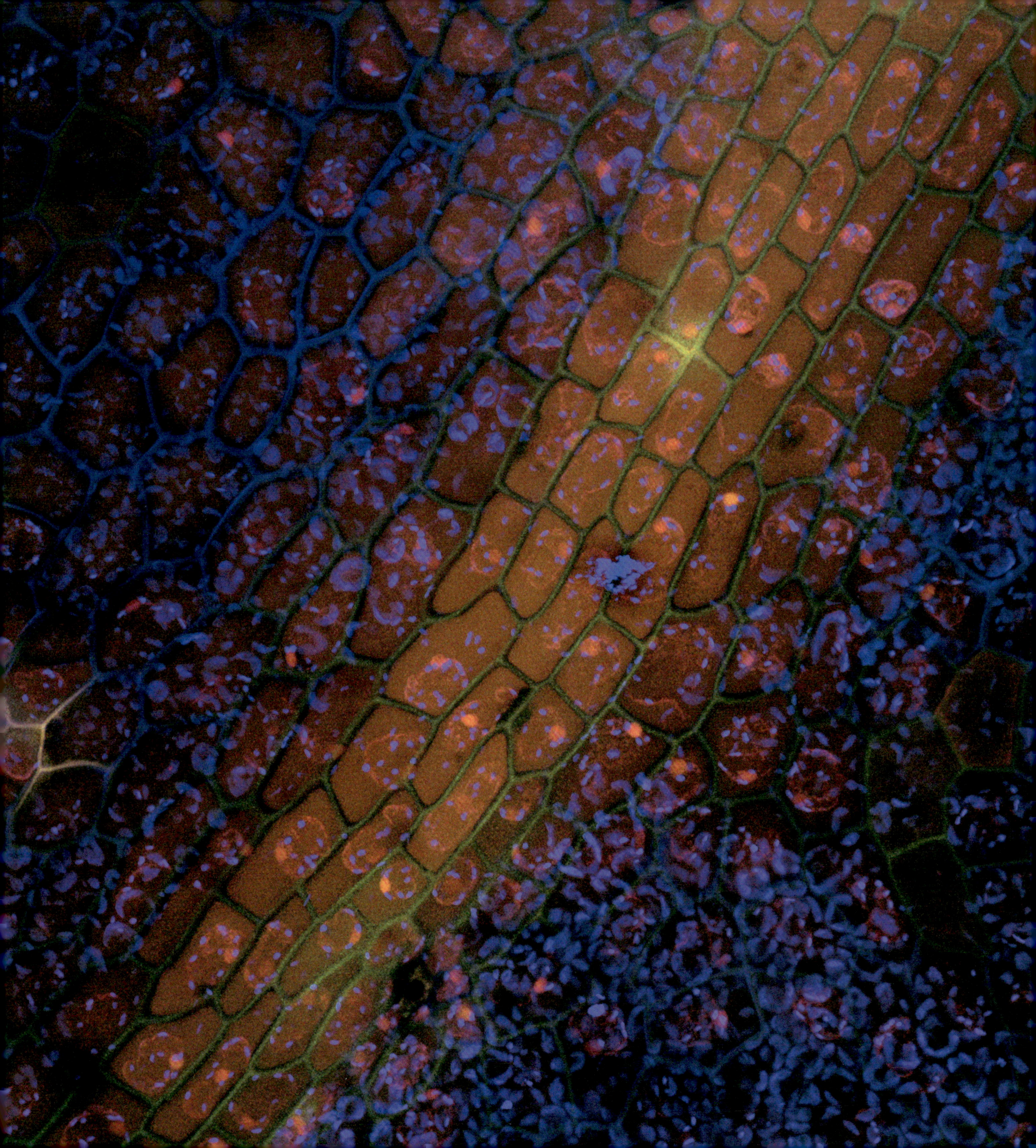

above: *Diplopterys longialata* (huambisa); flower
opposite: *Diplopterys cabrerana* (chaliponga); confocal image
50 μm

overleaf left: *Psychotria carthagenensis* (amyruca); confocal image
50 μm

overleaf right: *Psychotria viridis* (chacruna); confocal image
50 μm

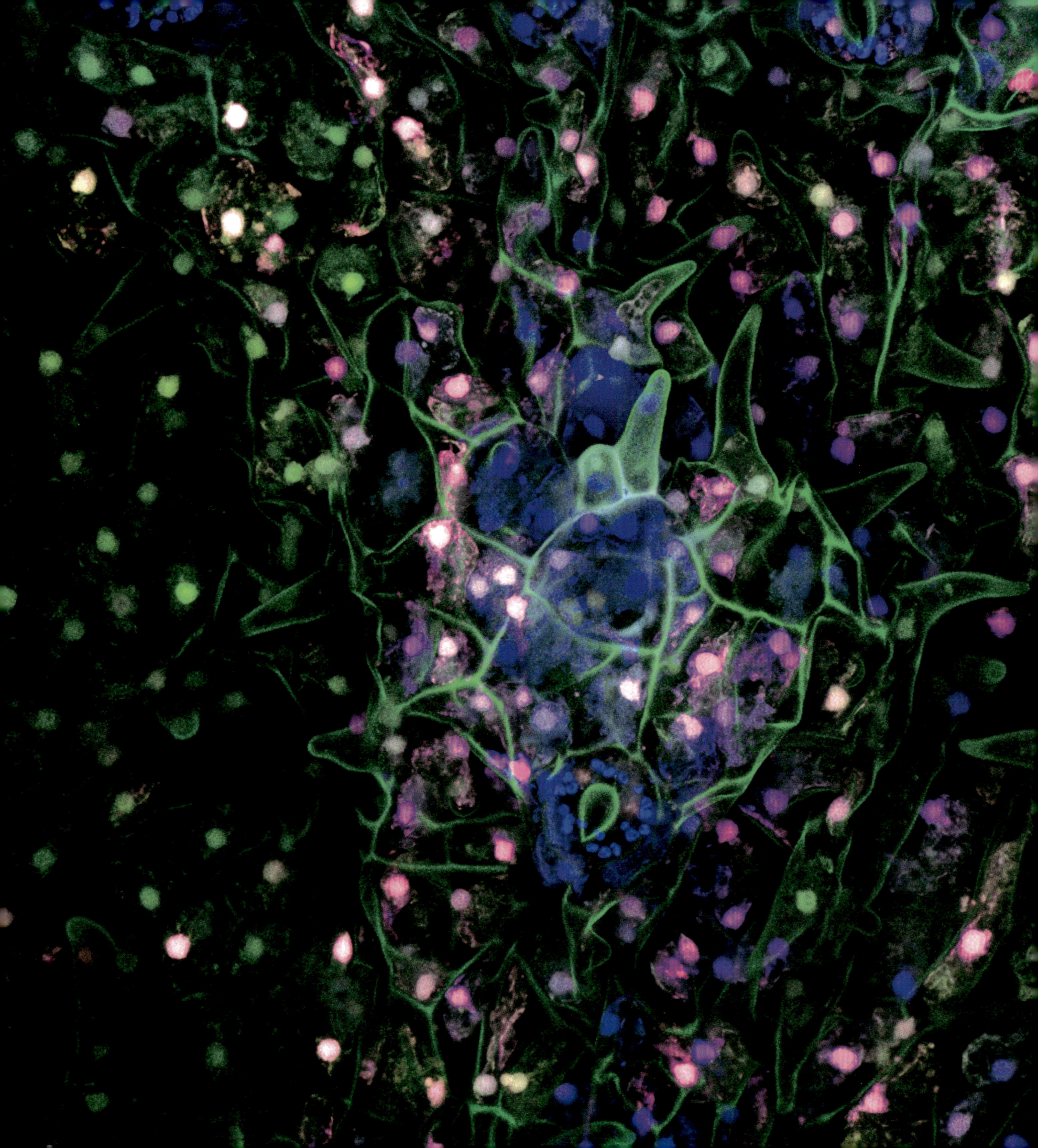

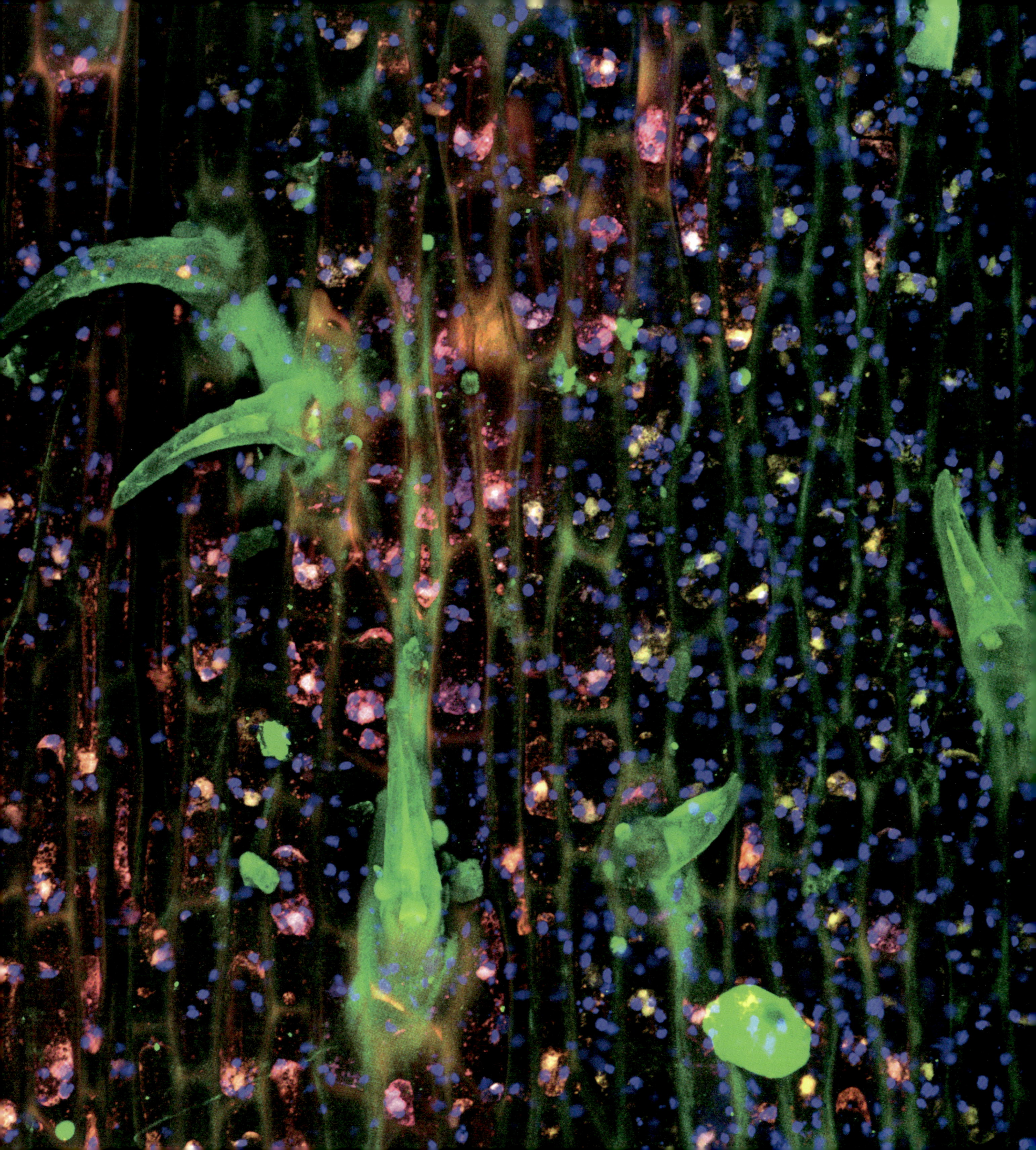

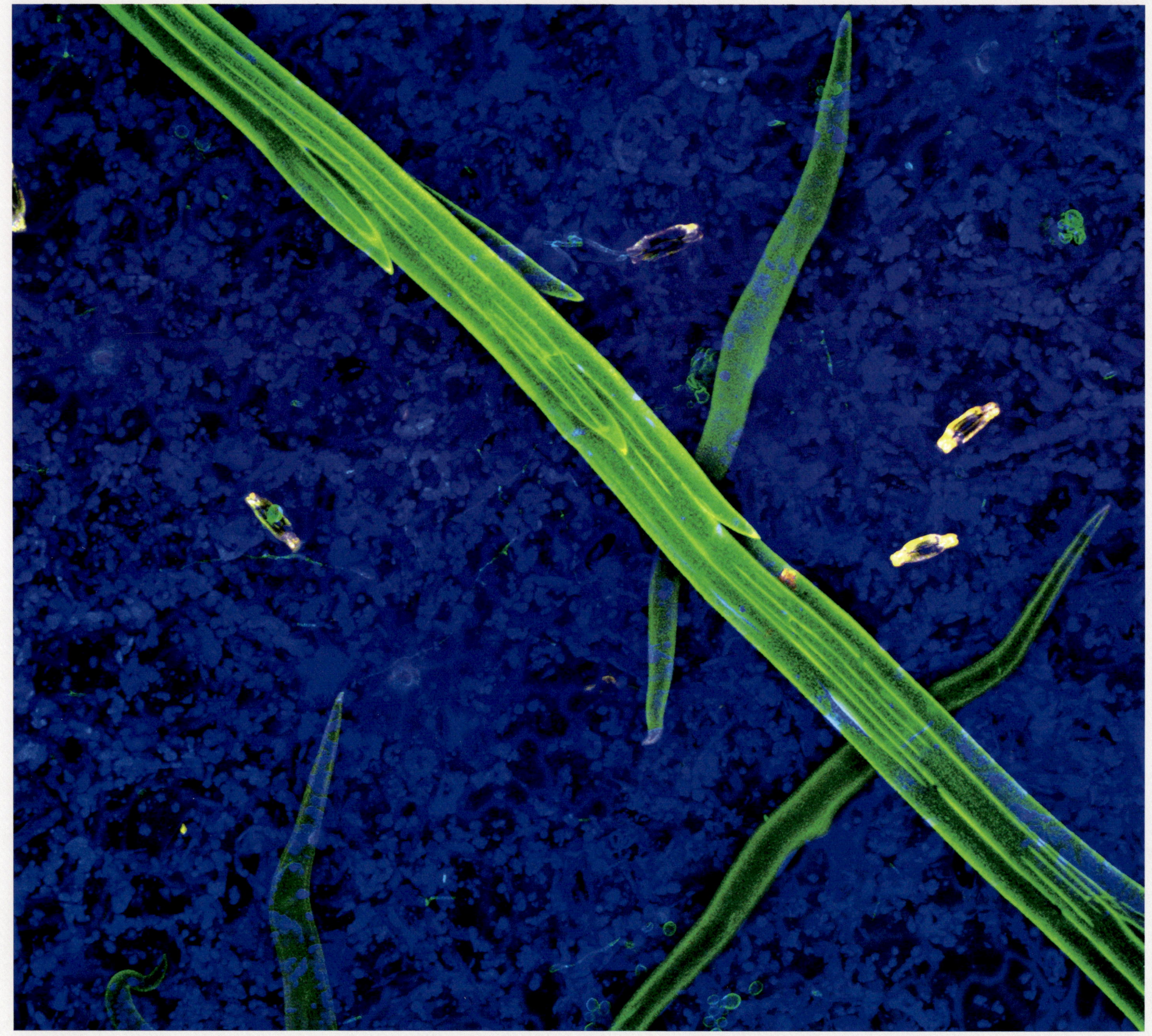

above: *Banisteriopsis caapi* (ayahuasca / yagé); confocal image

50 μm

opposite: *Banisteriopsis muricata* (red ayahuasca); confocal image

50 μm

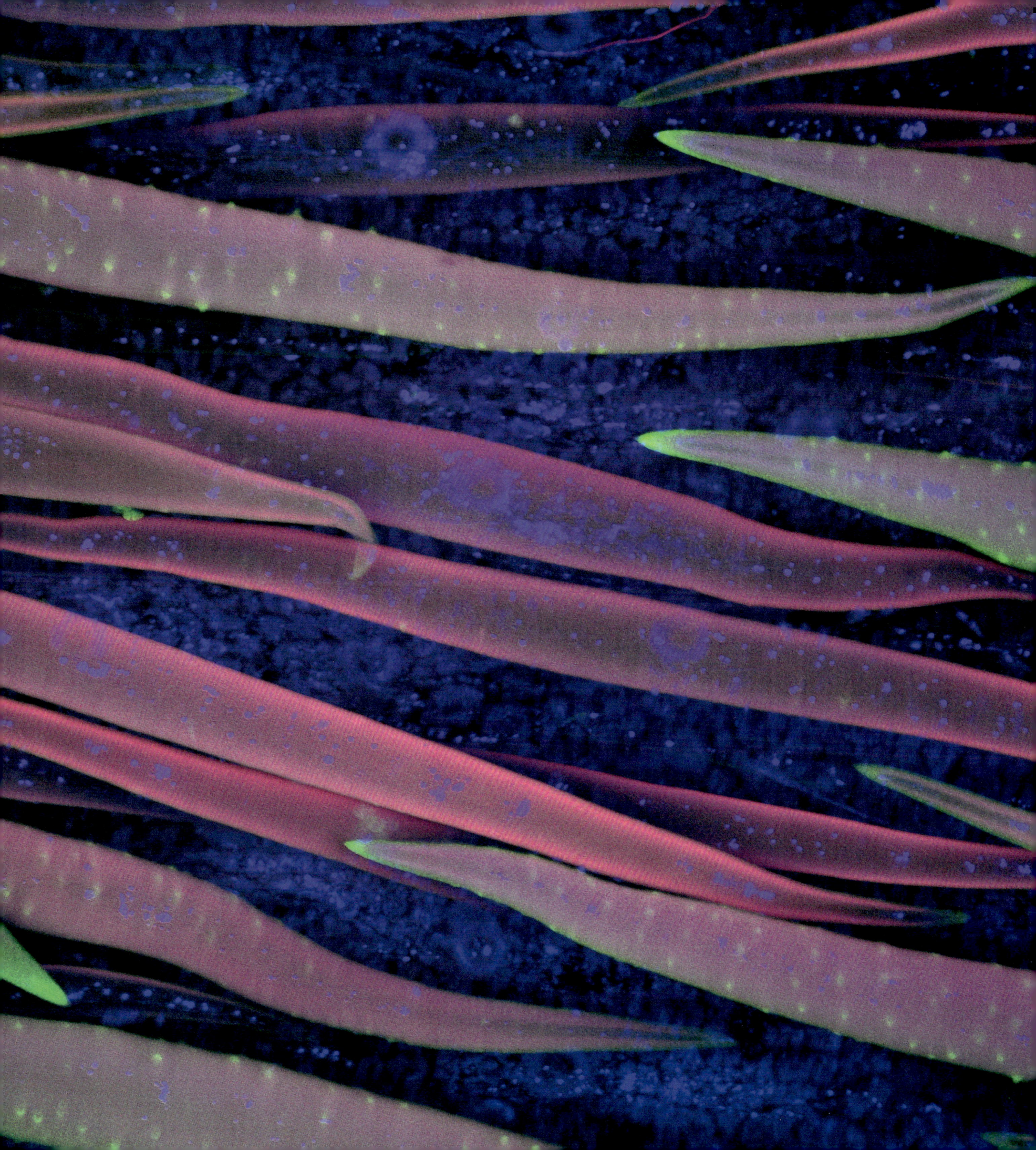

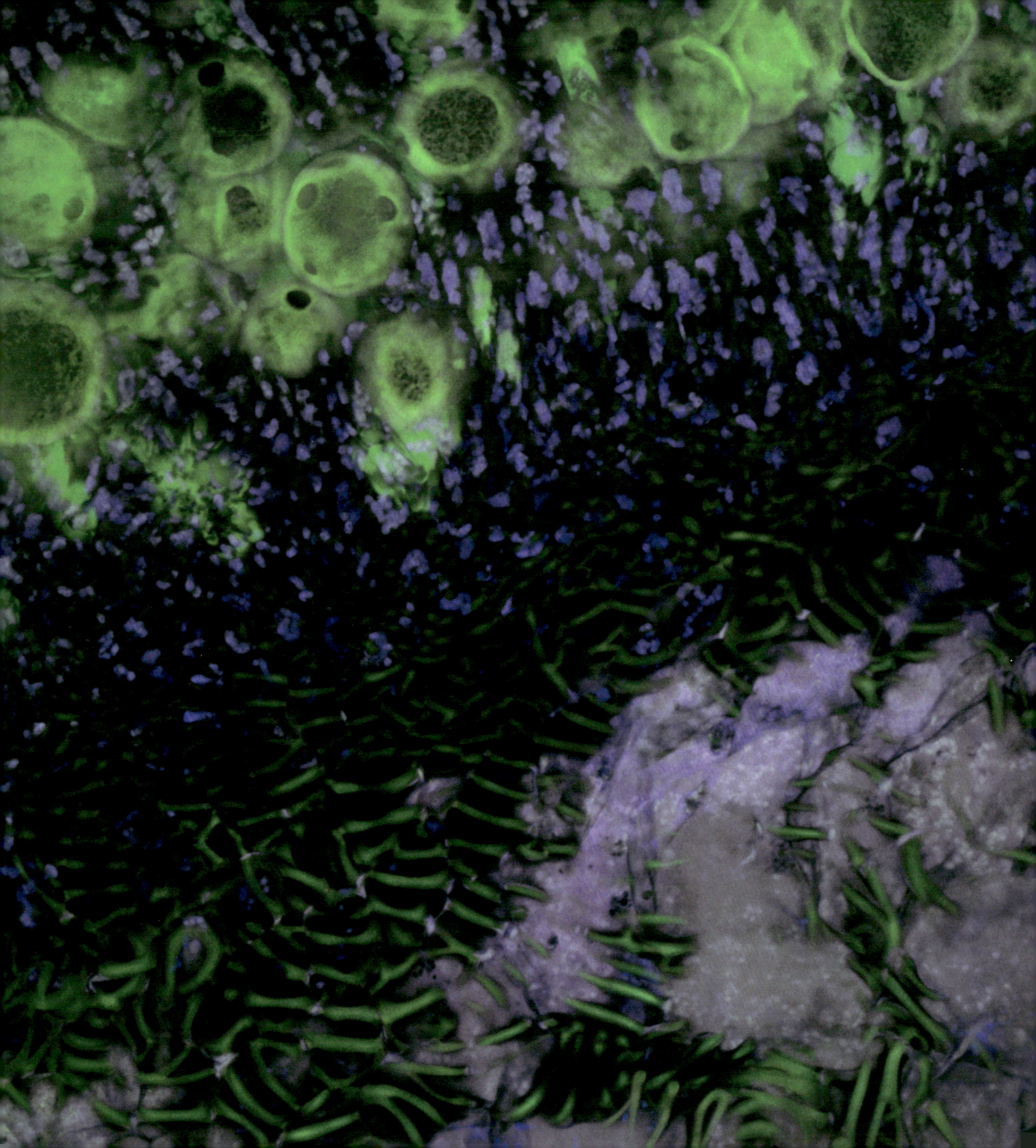

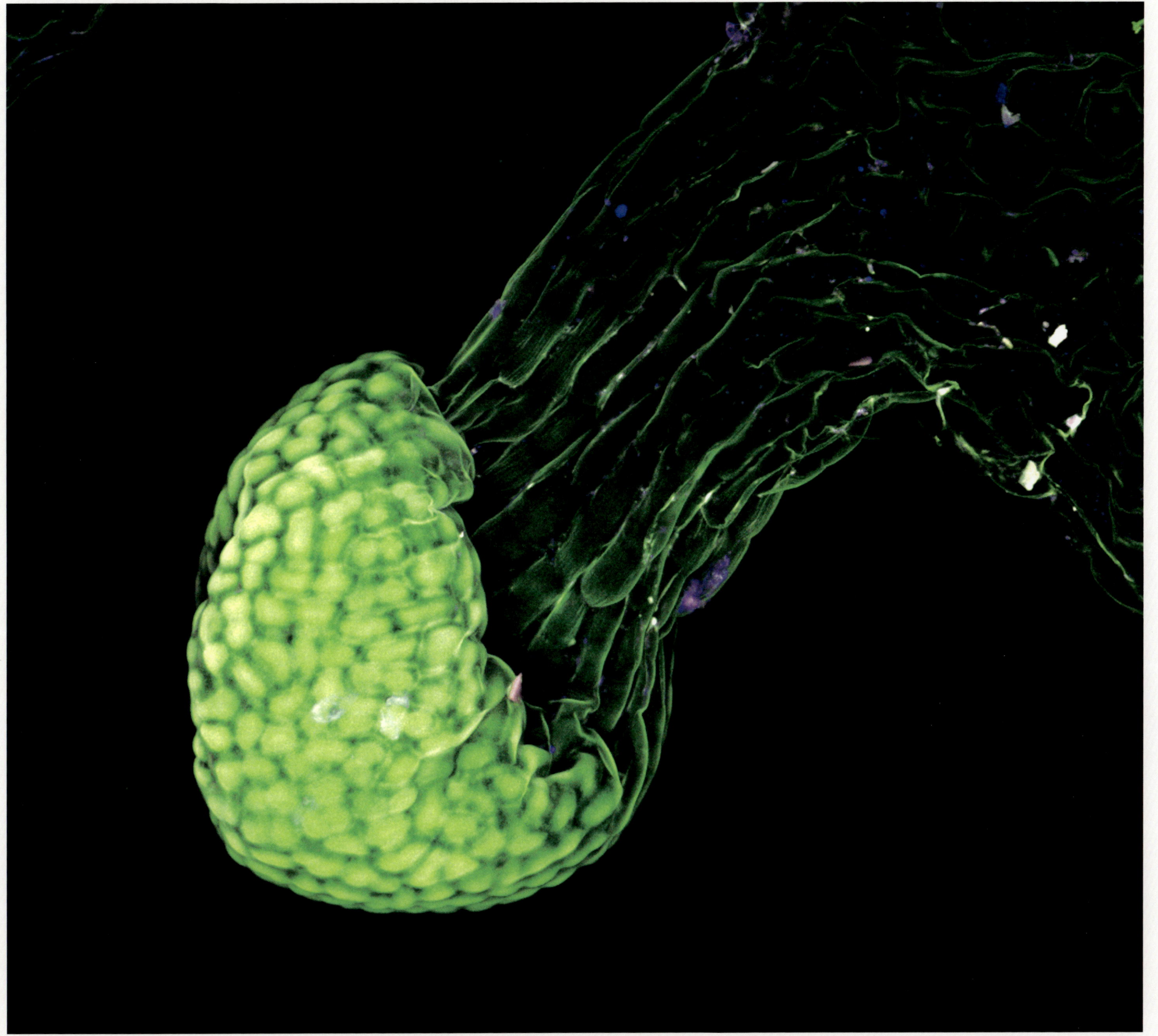

above: *Banisteriopsis caapi* (ayahuasca / yagé); confocal image

50 μm

opposite: *Banisteriopsis caapi* (ayahuasca / yagé); confocal image

50 μm

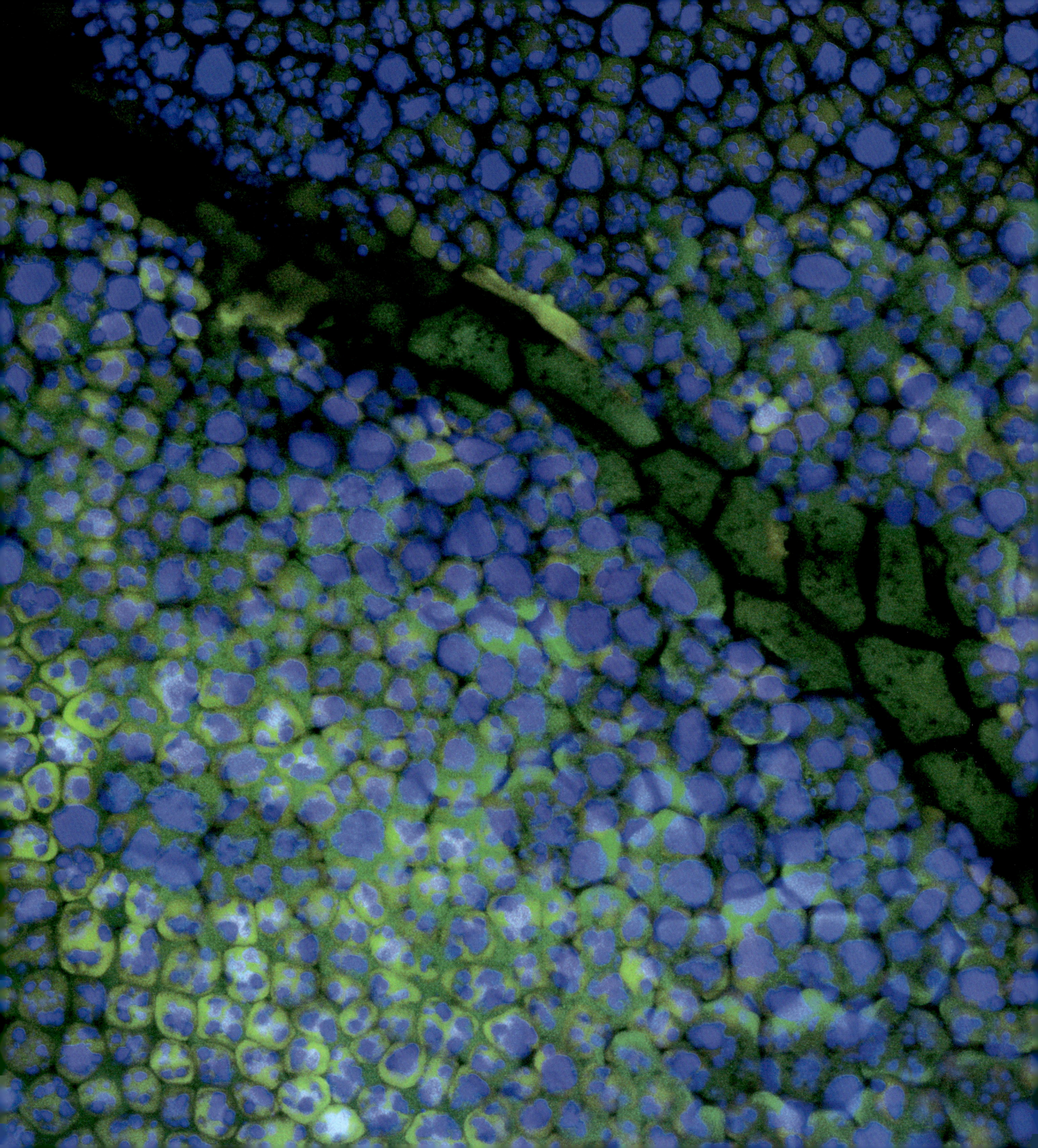

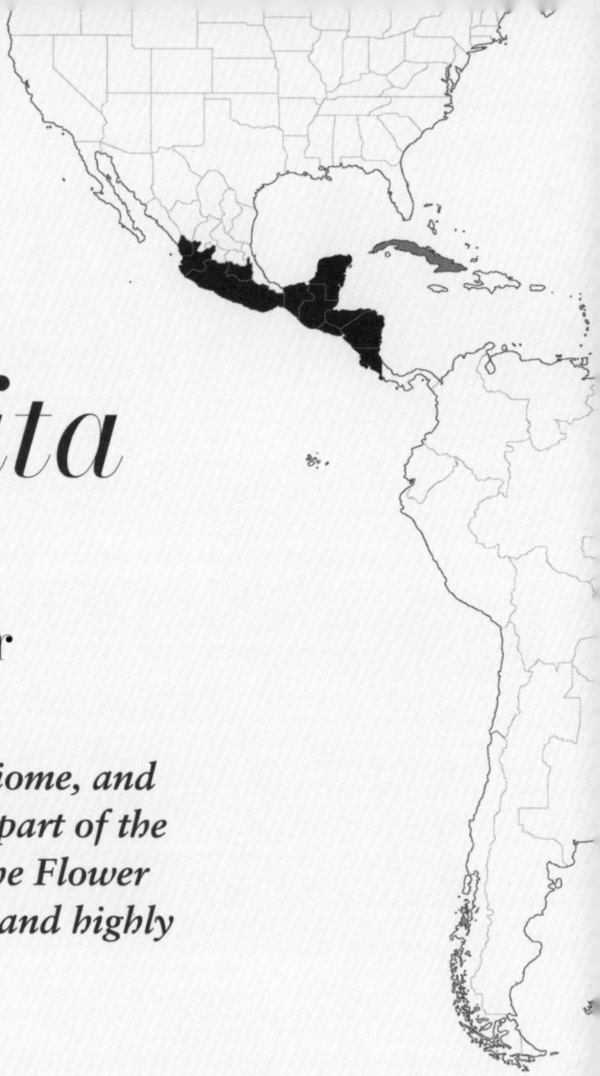

Boraginaceae

Bourreria huanita

Esquisúchil, guie xoba, ik'al te, jazmín de Oaxaca, popcorn flower

This Mesoamerican evergreen tree prefers a wet tropical biome, and was used in the tombs of Maya royalty as a quintessential part of the olfactory landscape, the now-forgotten perfume-path to the Flower World of the next life. It is extremely difficult to propagate and highly endangered by climate change.

An extremely compelling example of archaeobotanical research can be appreciated in Cameron L. McNeil's "The Flowery Mountains of Copan: Pollen Remains from Maya Temples and Tombs", in which she describes how she collected samples and undertook pollen analysis to work with microscopic evidence as a way to positively identify four plants used ritually at this particular ancient Indigenous site in Honduras: *Zea mays* (maize), *Typha* (cattails), *Acrocomia aculeata* (coyol palm) and *Bourreria huanita*, which, together, also constitute valuable olfactory information in this sacred context. This residue of these plants was contained in the burial chambers of the Rosalila Temple. *Bourreria* is especially significant in that, according to McNeil, "its lovely, sweet-smelling, yellow-centred white flowers would have imbued the buildings with a powerful paradisiacal fragrance", a species capable of "possibly channelling the breath soul of the deceased ancestral parents of the polity, or perhaps ushering them on to their flowery paradise". In her exemplary study, McNeil laments a sad reality: "various scholars have written about the flowers probably used by the Maya in the pre-Columbian period, but no one has analysed the microbotanical remains from temple and tomb floors to determine exactly what flowers actually had a role in rituals.

Bourreria huanita (esquisúchil); confocal image

50 μm

above: Esthela Calderón collecting two leaves and a dried flower from a rare *Bourreria huanita* (esquisúchil) inside the San Miguel Cathedral in Tegucigalpa, Honduras, 2022
right: *Bourreria huanita* (esquisúchil) flowers, Oaxaca, Mexico
opposite: Maya ruins of Copán, Honduras, site of innovative research involving archaeobotanical pollen analysis

This is unfortunate; archaeologists have swept away a wealth of information on ancient plant use as they uncovered the stone, stucco, and dirt of ritual spaces". The innovative and inspiring pollen analysis research conducted by McNeil indicates that "the white fragrant flowers of this tree were associated with the dead and used by the Mexica of highland Mexico as offerings in temples, in sacred gardens, as medicine, in cacao beverages, and as garlands to adorn individuals for sacred rites". She goes on to characterise ancient use of this species in the following way: "*Bourreria* is also tied to blood in many of its medicinal uses, and the plant may have been used to heal the wounds of autosacrifice". In addition, McNeil also asserts: "current evidence indicates that *Bourreria* is the most important ritual flower of the Maya to have been forgotten or lost in the time since the conquest".

After receiving permission to do so, Esthela Calderón collected two leaves for the *Microcosms* project from a *Bourreria huanita* tree (one of perhaps two in all of Honduras) located in an inner patio in the San Miguel Cathedral in Tegucigalpa, Honduras. During her visit, she spoke to the people whose job it is to clean the enclosed area where this ancient Esquisúchil is protected. They told her that they won't sweep the area in the very early hours at dawn during the months when the tree is in bloom because the intense fragrance of the flowers causes a kind of otherworldly lightheadedness. Thanks to conservationists, more of these threatened trees survive in Guatemala, especially in the cooler climate of Ciudad Antigua.

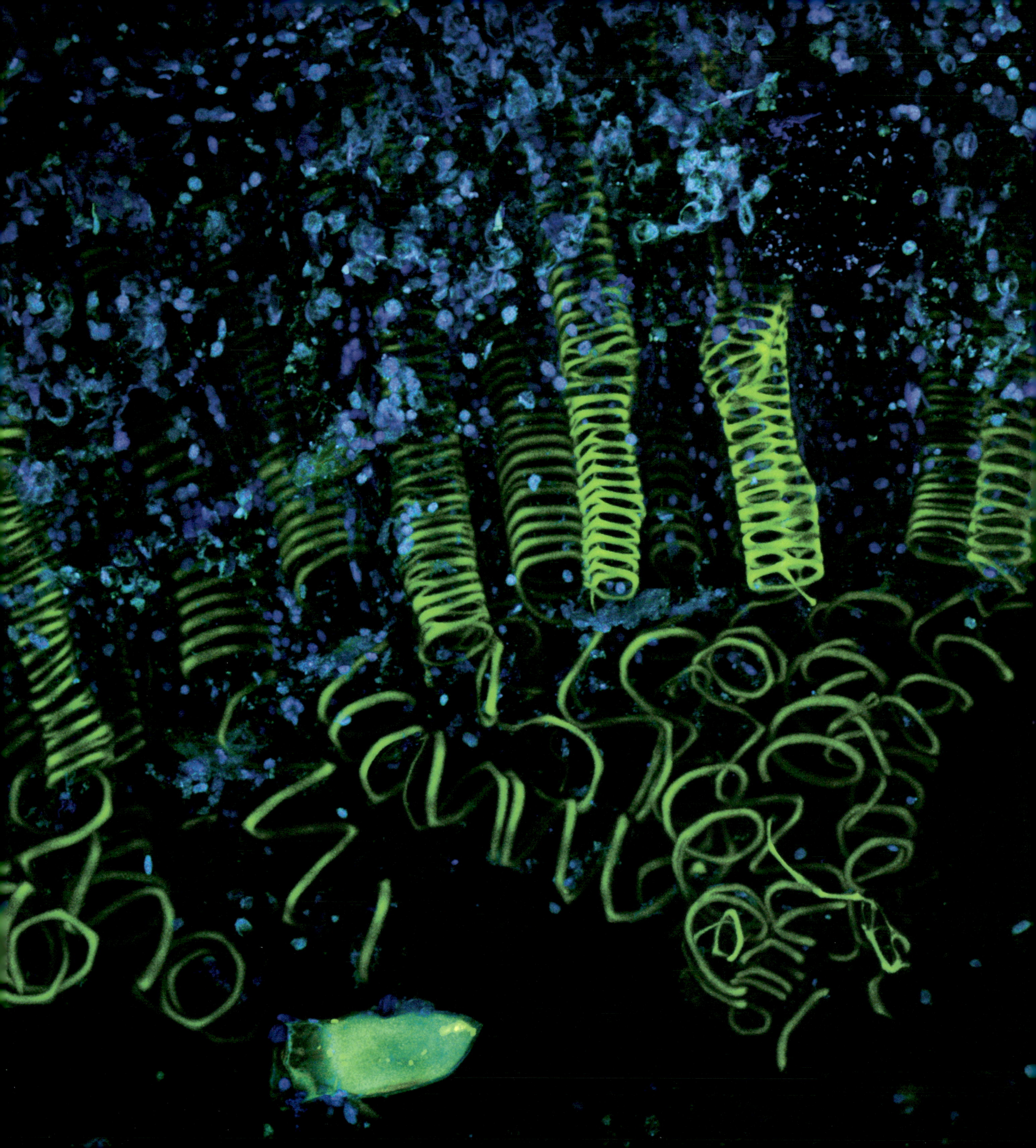

Solanaceae

Brugmansia insignis

Angel's trumpet, floripondio, huanduj, pehí, toé

This heavily-branching tree that prefers the cooler temperatures of the Andean highlands has a multitude of trumpet-shaped flowers that hang straight down and exude a bewitching nocturnal fragrance. Brugmansias are known as cultigens, and not plants that grow in the wild. They are propagated by human beings most readily by cuttings and are often placed at strategic points around dwellings for protection from malevolent spirits.

Brugmansia × candida

Culebra borrachero, lengua de tigre, mutscuai borrachero

This very distinctive and highly-revered Brugmansia hybrid cultivar created by Kamëntsá healers in southern Colombia's Sibundoy Valley has long snake-like leaves deformed by a virus.

Brugmansia sanguinea

Blood-red angel's trumpet, misha toro, puka wantuk

This tree grows in the montane tropical biome and has flowers that are greenish-yellow, bloodred, and trumpet-shaped that are more narrowly tubular than other Brugmansia species. It is cultivated commercially to produce scopolamine for relieving nausea associated with chemotherapy treatments and seasickness.

Brugmansia × candida (culebra borrachero)

50 μm

Whenever possible, *Microcosms – Sacred Plants of the Americas* seeks to highlight the potential connectivity between ancestral knowledge about medicinal plants and contemporary Western scientific models for the study of botany. In this sense, one of the most notable cross-cultural collaborations occurred when Kamentsá healer Salvador Chindoy shared his plant wisdom in the 1940s with Richard Evans Schultes. The renowned Harvard ethnobotanist hoped to learn more about the psychoactive and healing properties of the rare Brugmansias that grow only in Colombia's Sibundoy Valley. Now, many decades later, Bernardo Chindoy, who has preserved what he learned from his grandfather Salvador and both his parents, is working with Federico Roda, a professor at the Universidad Nacional de Colombia (Bogotá), who specialises in the evolutionary genomics and metabolisms of the same Brugmansias that so intrigued Schultes. Together, they will share paradigms for a fuller understanding of these extremely potent members of the Solanaceae family, mysterious cultivars with names like Amarón, Biangán, Buyés, Culebra, Dientes, and Quinde. In conversations with Roda, *taita* Bernardo has spoken of the crucial importance of creating on his family's land a garden of these medicinal Brugmansias, also known collectively as *borracheros*. He mentioned that when he was younger, he did indeed cultivate and conserve just such a garden. Tragically, however, the US government's entirely misguided War on Drugs with its catastrophic ecological consequences destroyed the shamanic plants thriving under his skilled care. He watched as the revered vegetal beings that he used to heal his patients burned and shriveled after the aerial spraying of herbicides containing glyphosate. Bernardo Chindoy and Federico Roda are hopeful that a new beginning is possible and that the international community will donate to this worthy cause.

Alistair Hay, one of the authors of the exquisite book *Huanduj: Brugmansia*, affirms that "the Incan people were relative newcomers to the Peruvian scene, bringing diverse Indigenous groups under their domination, and they came late in their history to embrace the sacred plants which had long been a central part of the religious cultures of those they conquered: the hallucinogenic brugmansias were among the most important to them".

He goes on to say that brugmansias "are without question the elite South American entheogen, usually reserved for the ultimate in shamanic training, the most difficult cases of divination and healing, for the fiercest of warriors and the most courageous and skilled of shamans".

Confirming this idea, anthropologist Glenn H. Shepard, Jr. maintains that, according to what he has observed in the Peruvian Amazon, "a larger, vision-inducing dose of *Brugmansia* infusion may be given orally as a last resort to treat people with incurable illnesses, witchcraft, or severe accidents. This preparation is considered to be the most intoxicating (*kepigari*) and strongest of all medicines… *Brugmansia* is the open-heart surgery of the Matsigenka – a final resort to the highest medical authority, reserved only for the most drastic cases".

An article by Arteaga de García from the specialised academic research publication *Revista Colombiana de Ciencias Químico-Farmacéuticas* posits Brugmansias as a promising species for the production of tropane alkaloids, especially scopolamine and atropine, which are used in anticholinergic drugs to treat chronic obstructive pulmonary disorder and

top left: Kamëntsá traditional healer Taita Salvador Chindoy, Sibundoy Valley, Colombia, 1942
above left: Taita Bernardo Chindoy, grandson of Salvador, Medellín, Colombia, 2025
above right: Taita Salvador Chindoy with ethnobotanist Richard Evans Schultes and translator Pedro Juajibioy, Sibundoy Valley, Colombia, 1942

overactive bladder as well as to alleviate nausea from motion sickness and chemotherapy. This research revealed that the leaves and flowers of *Brugmansia sanguinea* were found to have especially high concentrations of these compounds.

Current research in Brazil led by Sandro Pinheiro da Costa adopts a global perspective on the therapeutic uses of *Brugmansia suaveolens*, citing by country where the preparations made with different parts of the plant are used to treat asthma and bronchial problems, gastric disorders, infections, wounds, ulcers, rheumatic pain and vaginal infections. There is also a section dedicated to toxicity related to Brugmansias in the form of hallucinations, hysteria, and other anticholinergic symptoms. Included in this study is a brief analysis of the essential oils in *Brugmansia suaveolens* flowers that vary as the blossoms mature and change colour from yellow to white to pink over a 24 hour period.

Painting of *Brugmansia suaveolens* (angel's trumpet) and *Brugmansia sanguinea* (blood-red angel's trumpet); oil on canvas, by Peruvian artist Anderson Debernardi

above: *Brugmansia sanguinea* (blood-red angel's trumpet) at the Ingapirca ruins, Ecuador
overleaf left, top: *Brugmansia* × *candida*; (culebra borrachero)
overleaf left, bottom left: *Brugmansia* × *candida*; (culebra borrachero)
overleaf left, bottom right: *Brugmansia sanguinea* (blood-red angel's trumpet)
overleaf right: *Brugmansia insignis* (angel's trumpet)

Microcosms – Sacred Plants of the Americas

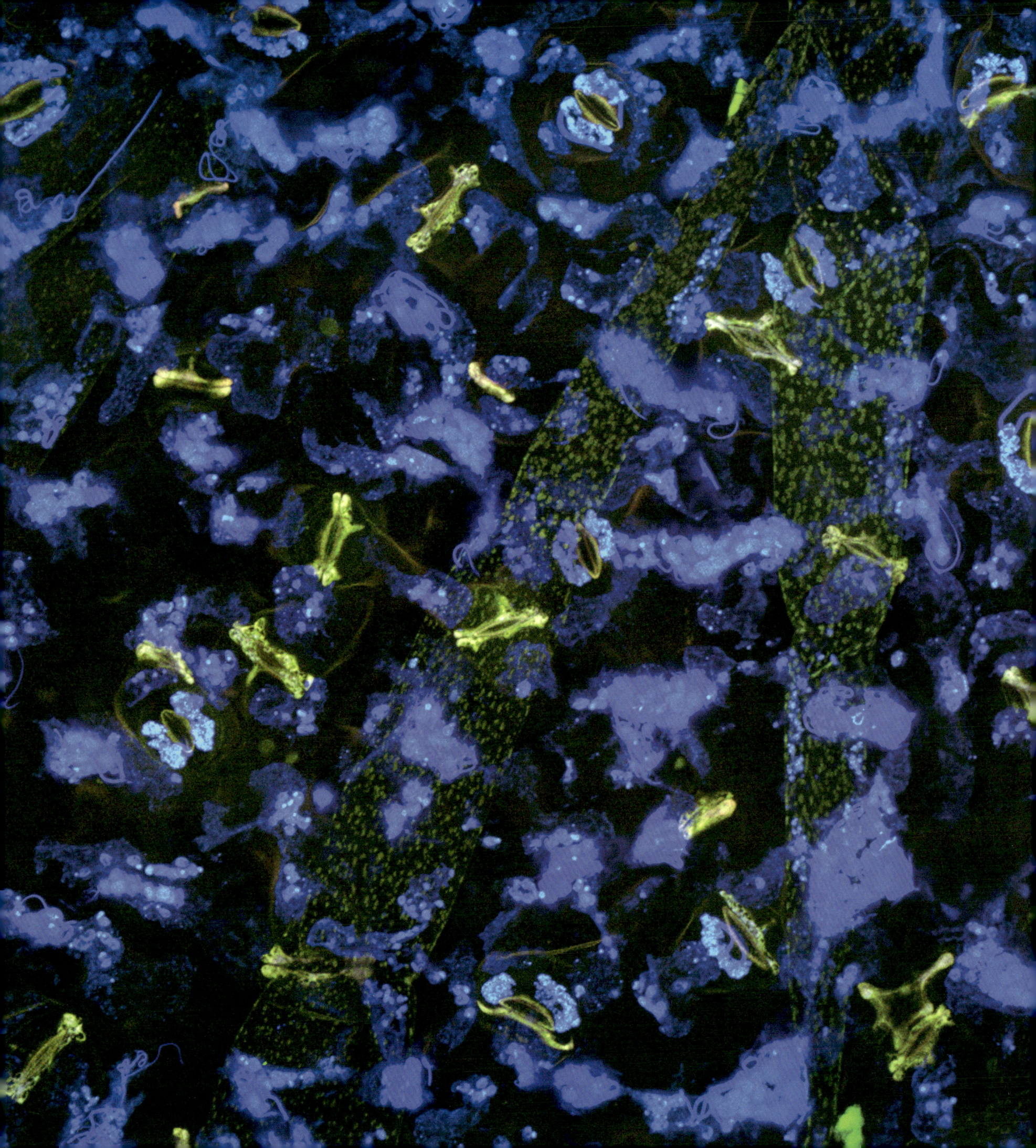

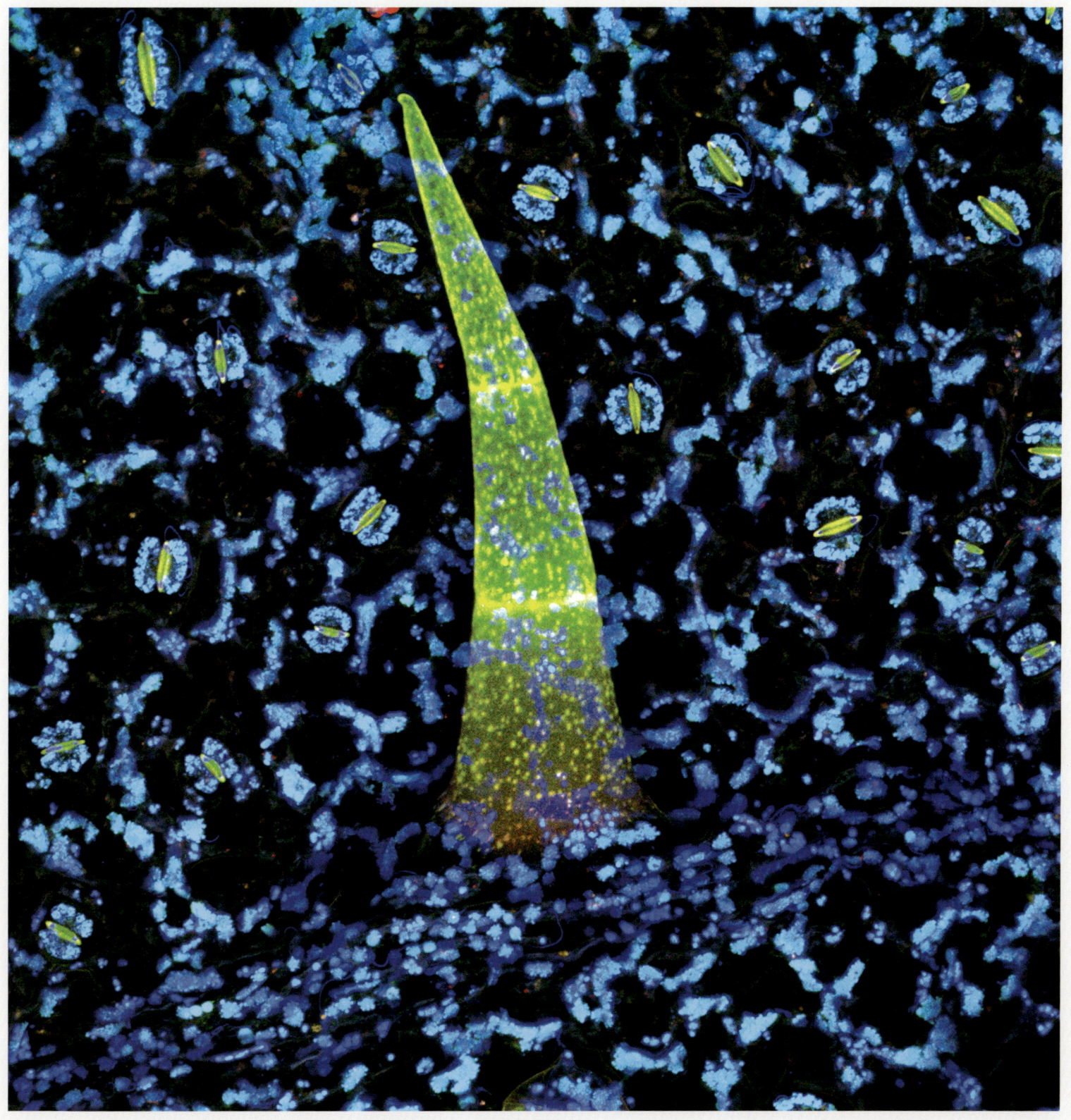

above: *Brugmansia insignis* (angel's trumpet); confocal image
50 μm

opposite: *Brugmansia sanguinea* (blood-red angel's trumpet); confocal image
50 μm

overleaf left and right: *Brugmansia sanguinea* (blood-red angel's trumpet); confocal images
50 μm

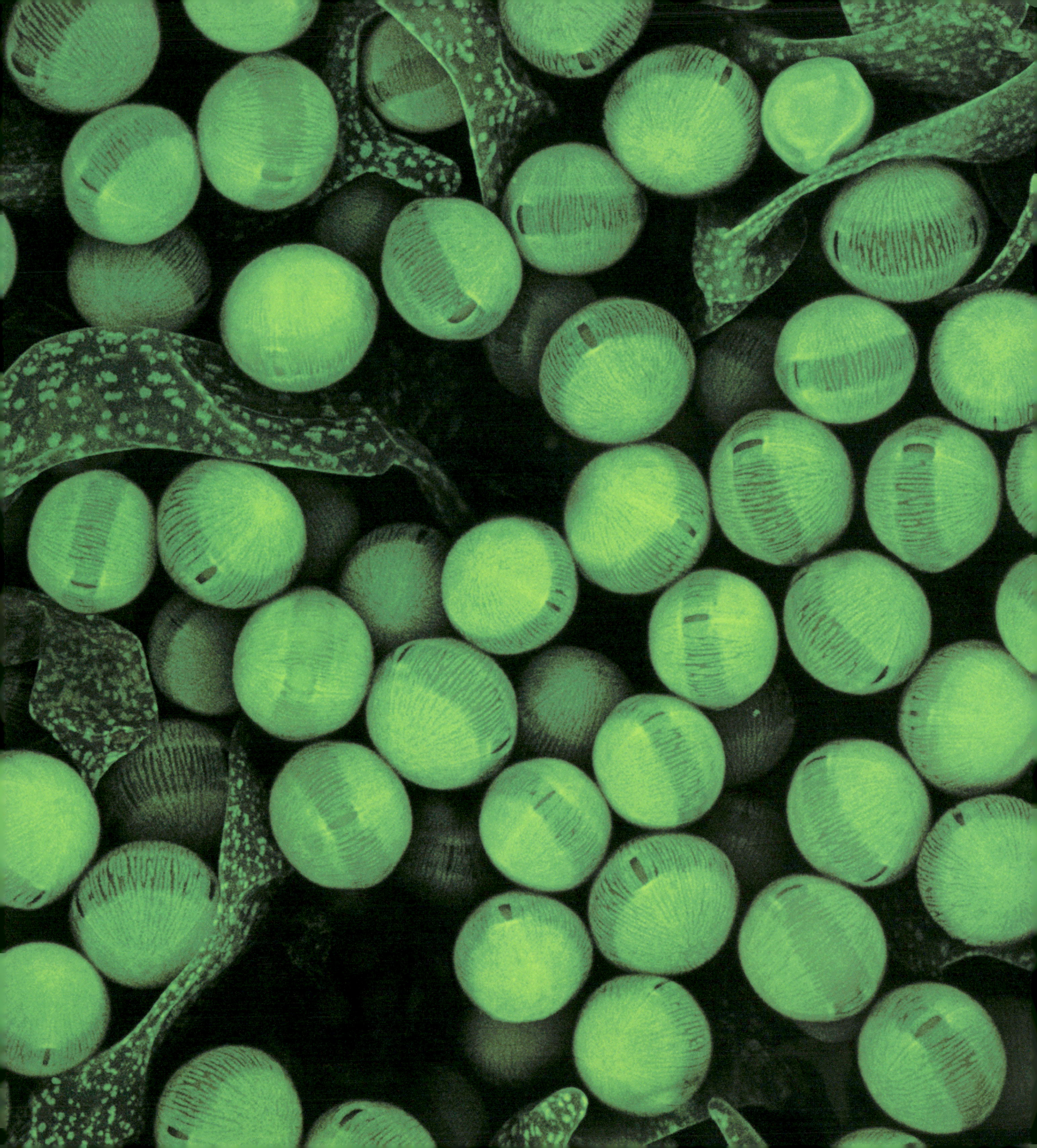

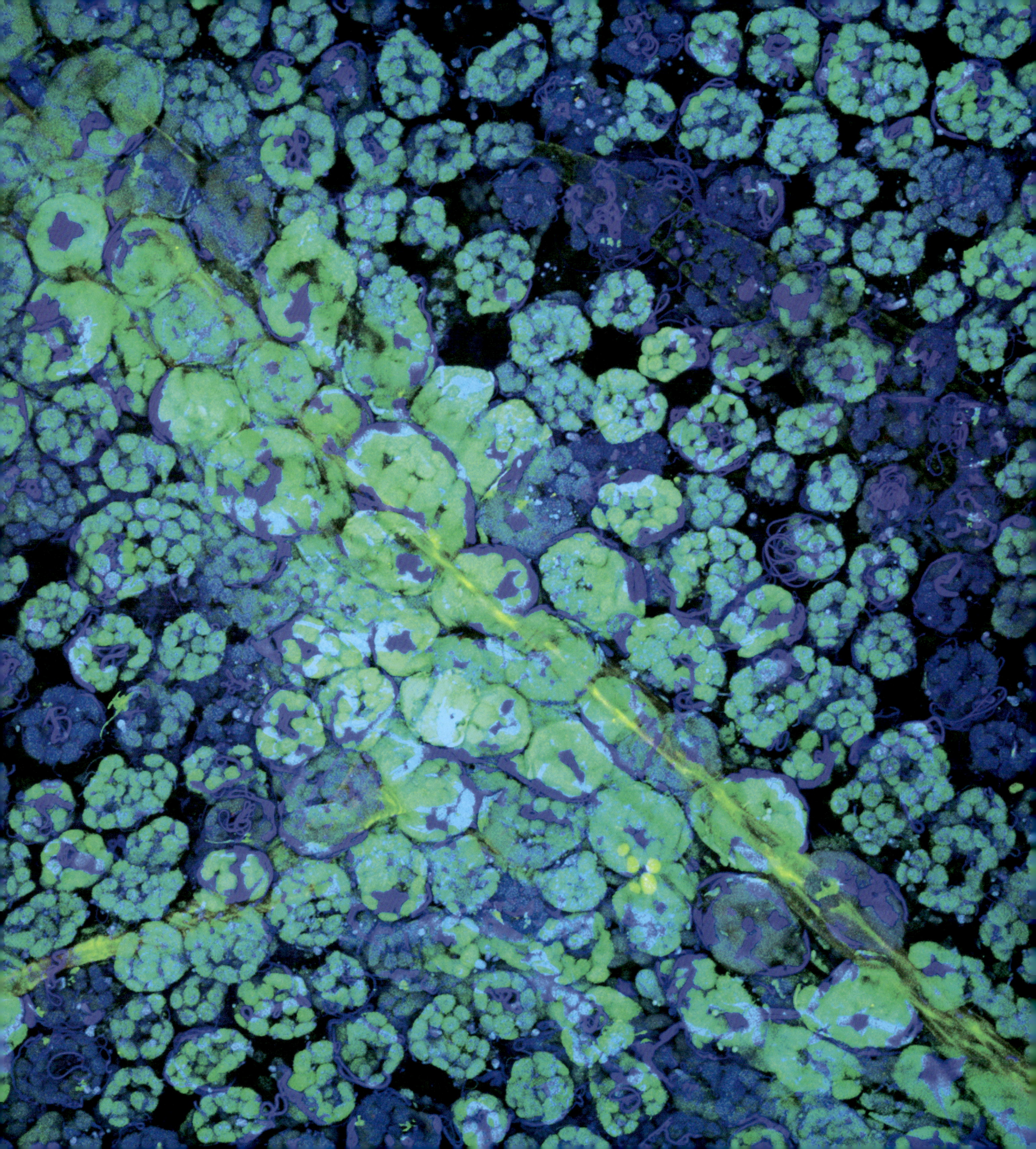

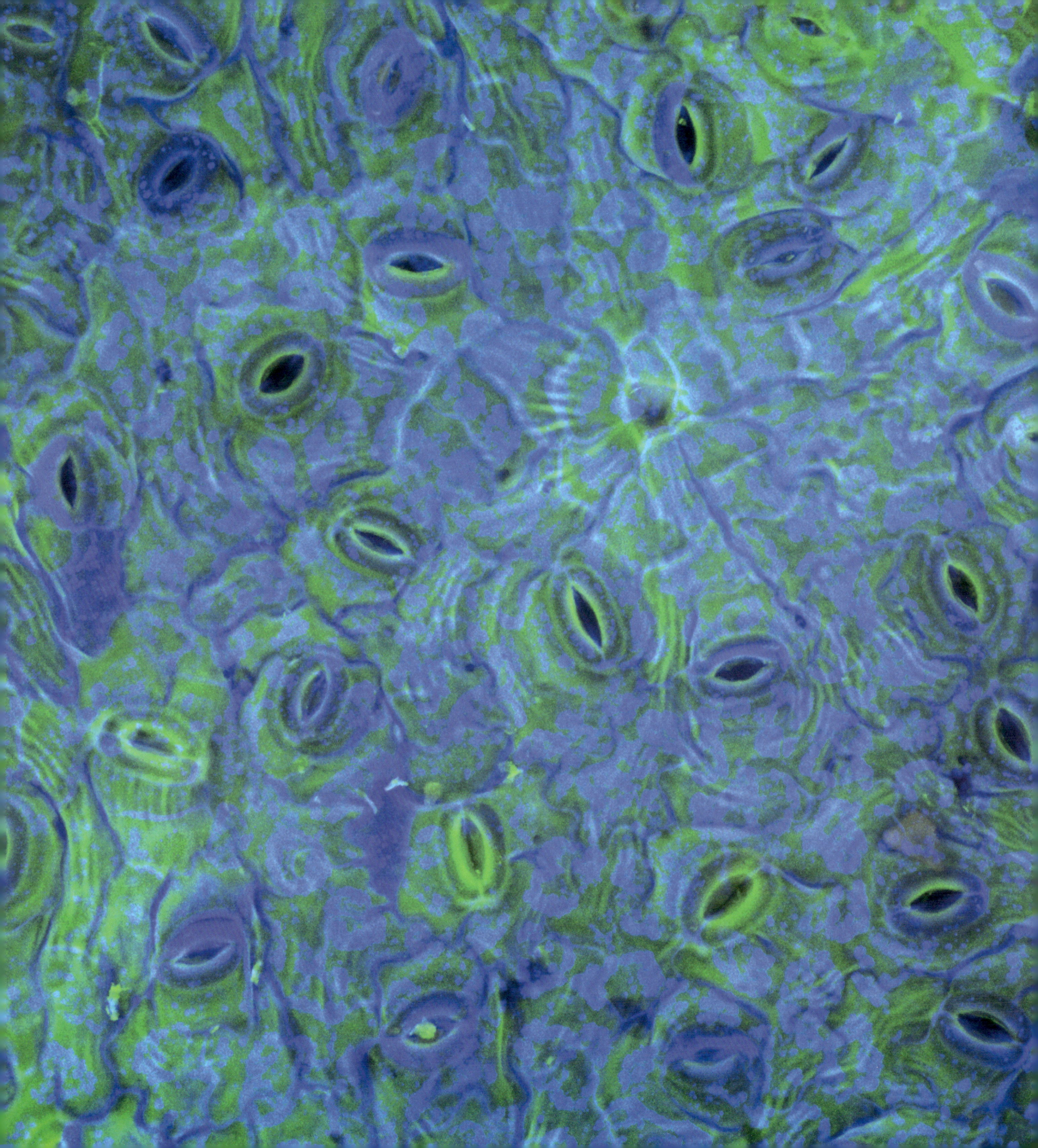

Solanaceae

Brunfelsia grandiflora

Chiric-sanango, chiri kaspi, ujajái, yesterday-today-tomorrow

This evergreen shrub grows in a tropical climate and has flowers that are dark violet in colour, which fade to lilac and then white. Traditional Amazonian healers have the highest regard for this powerful plant-teacher. The Canelo-Kichwa community in Ecuador uses this plant, known as chiric-sanango, in the medical areas related to childbirth and anesthaesiology.

The stomata are the oval-shaped apertures in plants through which the carbon dioxide necessary for photosynthesis enters. They also release oxygen. In some of the confocal images of *Brunfelsia grandiflora* included here, the stomata are especially evident, and could serve as an opportunity to consider how breath links humans and plants. In keeping with the ideas of Monica Gagliano in *Thus Spoke the Plant*, with every breath, plants in our presence might know humans as themselves, and, reciprocally, we humans become more plant-like than we realise and perhaps are able to recognise how we, too, can know plants as ourselves.

The flowers of *Brunfelsia grandiflora* are a dark purple the first day, a light lilac colour the second, and very white the third. The common name in English for this prized ornamental, "yesterday-today-tomorrow", is an invitation to meditate on the nature of Time itself, and how temporality with its fluid, ongoing cycles is fully contained in each and every living organism.

Brunfelsia grandiflora (chiric-sanango); confocal image

50 μm

As a prelude to mentioning *Brunfelsia grandiflora* and how it is perceived by different Amazonian ethnic groups and neighbouring mestizo populations, Ludmila Skrabáková, who worked in the Peruvian Amazon among the Shipibo and Ocaina from 2002-2010, defines the key concept of *Amerindian Perspectivism*: "together with other subjects, plants are entities that act as humans and thus as social agents – not just in traditional Amazonian healing and magic, but also in everyday life. The spirits or souls of plants, called *madres* (mothers) or *dueños* (owners), have anthropomorphic traits and are holders of certain characteristics, qualities, and powers. They are respected and feared; they intervene in peoples' lives whether as a means of granting health or wealth or as teachers, guardians, villains, and evildoers, and they occupy a solid place in Amazonian cosmology and mythology".

In *Las visiones y los mundos*, Pedro Favaron says that the theory of Amazonian Perspectivism, developed by Brazilian anthropologist Eduardo Viveiros de Castro, affirms that "living is thinking and that goes for all living beings: from the smallest organisms to the Owners of medicine, from plants to humans; each species has its own perspective". Favaron also says that, accordingly, "relations between plants and humans should occur in terms of equilibrium and dialogue".

Skrabáková maintains that Amazonian shamans "are in constant negotiation with the plants' souls – they enter into contracts with them, call on them for assistance, learn from them, and mediate contact between their human fellows and plants". During her fieldwork, the researcher from the Czech Republic learned from these visionary healers that *Brunfelsia grandiflora*, or chiric-sanango, is "one of the most powerful master plants and doctors". The shamans, she continues, know this plant as "a wise old man with white hair. He is a very important teacher. With proper 'dieting', the juice from its roots and bark has the capability to open the doors to the plants' world and make one see and understand the nature of plants as they really are (their human nature/qualities)".

Timothy Plowman's article "*Brunfelsia* in Ethnomedicine", despite its 1977 publication date, is still the best overview of this genus and its medicinal and ceremonial uses throughout the Amazon region. Plowman's fieldwork and research confirm the importance of *Brunfelsia grandiflora* as an analgesic and medicine against rheumatism and arthritis among the Kokama from the Río Ucayali in the Peruvian Amazon, the lowland Quechuas of Ecuador's Río Napo, and the Siona of the Colombian Putumayo. As an admixture to the ayahuasca/yagé preparation among the Jívaro, Lama, Siona, Kofán, and Inga, Plowman speculates that *Brunfelsia* and the tingling sensations that it produces when ingested creates "striking tactile hallucinations" and serves "to create a greater physical awareness during the ceremony". Such a ceremony must be led by an Indigenous shaman highly skilled with regard to the dosage of this potent member of the Solanaceae family, given that it is also used as a piscicide, an arrow poison, as well as an antidote to snakebites.

In *One River*, Wade Davis recounts the story that Timothy Plowman told him about a near-death experience he had as a result of ingesting an extract prepared by a Colombian shaman. Davis writes: "only in this case the sensation grew to a maddening intensity, spreading from the lips and fingertips toward the centre of the body, progressing up the spine to the base of the skull in waves of cold that flooded over his consciousness. His breathing collapsed. Dizzy with vertigo, he lost all muscular control and fell to the mud floor of the shaman's hut. In horror, he realised that he was frothing at the mouth. An hour passed. Paralysed and tormented by an excruciating pain in his stomach, he remained only vaguely aware of where he was

"Amazonian shamans consider Brunfelsia grandiflora, or chiric-sanango, one of the most powerful master plants and doctors, and know this plant as a wise old man with white hair and a very important teacher."

Brunfelsia grandiflora (chiric-sanango); confocal image

50 μm

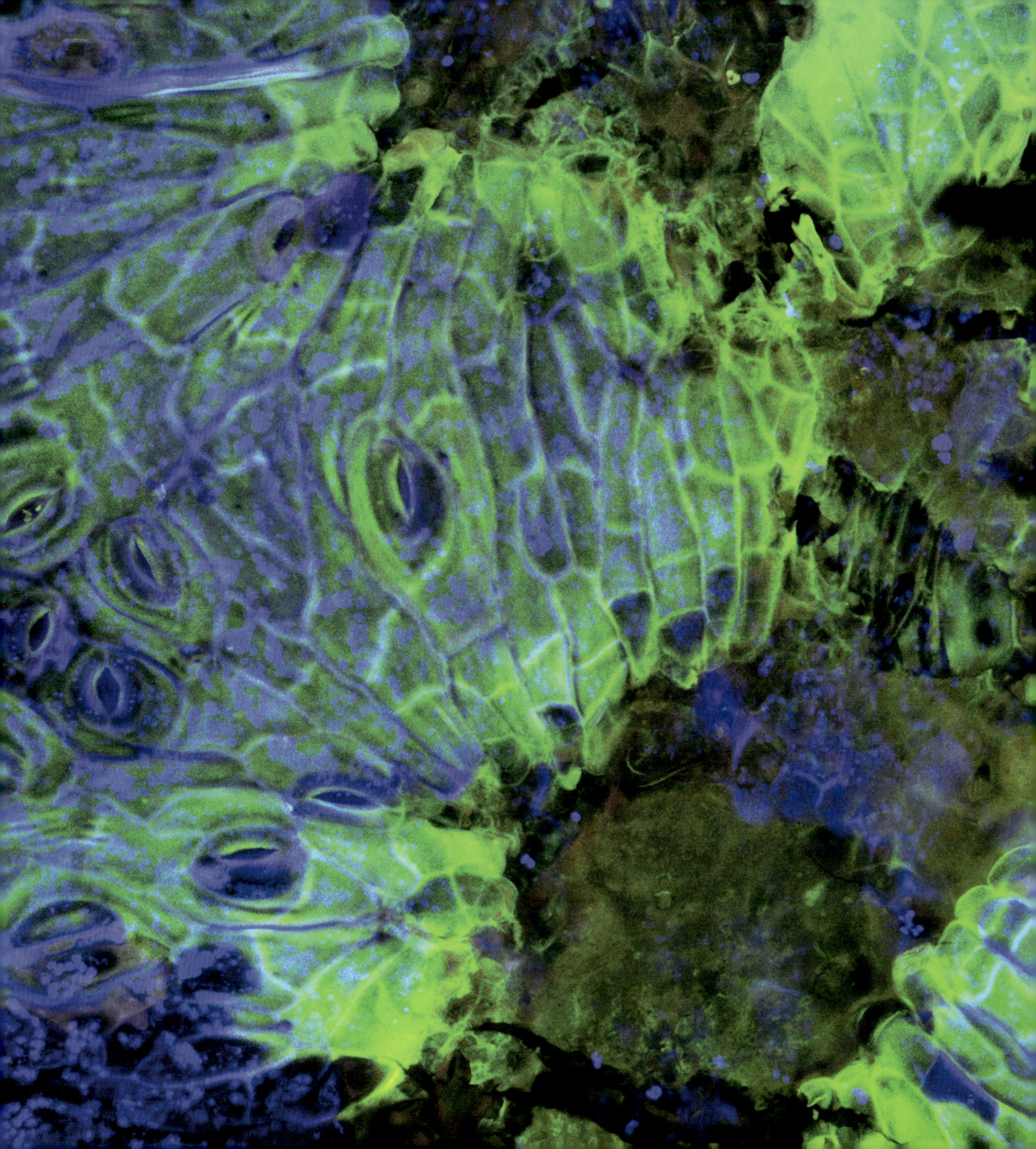

top: *Brunfelsia grandiflora* (chiric-sanango); flower

above: "Song of Chiric-Sanango", embroidered cloth, anonymous Shipiba artist, Peru

opposite: *Brunfelsia grandiflora* (chiric-sanango); confocal image

50 μm

Brunfelsia grandiflora

– on the earth, face-to-face with three snarling dogs fighting over the vomit that spread in a pool around his head".

A team of researchers led by Carmen X. Luzuriaga-Quichimbo from Ecuador's Universidad Tecnológica Equinoccial published a study in 2018 with a triple objective: 1) to synthesise the ethnobotanical knowledge about *Brunfelsia grandiflora* throughout Indigenous communities in Ecuador, 2) to rescue traditional knowledge about *Brunfelsia grandiflora* that is extant in a specific isolated Canelo-Kichwa Amazonian community in the province of Pastaza, Ecuador, and 3) to propose new bioproducts based on this plant related to the areas of childbirth, anaesthesiology, and neurology.

Raquel Mateos and her colleagues, in an overview of *Brunfelsia grandiflora* as a traditional medicine, identify for the first time "the phenolic composition of this medicinal plant to know the chemical structures of these phytochemicals that are behind the [plant's] renowned biological properties".

In a personal communication, Jonathon Miller Weisberger described his experiences with Ujajái (the name in the Paicoca language for *Brunfelsia grandiflora*) while he was living for an extended period of time with the Siekopai in Ecuador. He said the root of the plant is rasped and left as a cold water extract to "cook" in direct sunlight on the same day as yagé is being prepared. As time passes, the root starch of the *Brunfelsia*, which is poisonous, settles in the container of the extract. The maestro-healer drinks some pure yagé then adds only the liquid (not the accumulated solid) Ujajái to the yagé preparation, which the maestro stirs with a twig and prays over it for a good while. The maestro, having drunk yagé already, has entered the sacred ceremonial space, praying and blowing on the ujajái-yagé mix. From his own hand, he gives the one or two of his apprentices who are present at the ritual a gourd filled with the mixture. The students do not touch the gourd, but, rather, drink from the maestro's hand. Jonathon said that he drank this mixture on several different occasions and experienced similar effects, which he characterised as "agonising and maddeningly strong", and not something he looks forward to trying again: fire was burning up his body, shooting from his every orifice, and he saw two boas of fire like a caduceus burning upward from his belly. Other preparations that Jonathon mentioned involved ujajái leaves, and were used by the Siekopai for treating arthritic pain, and also toothaches.

Brunfelsia grandiflora, known by a plethora of Indigenous common names, is a plant of ancient traditions as well as a vibrant (albeit relatively unknown) life in the contemporary world of Amazonia. Clearly, it merits deeper study, and, most certainly, it is worthy of our utmost respect.

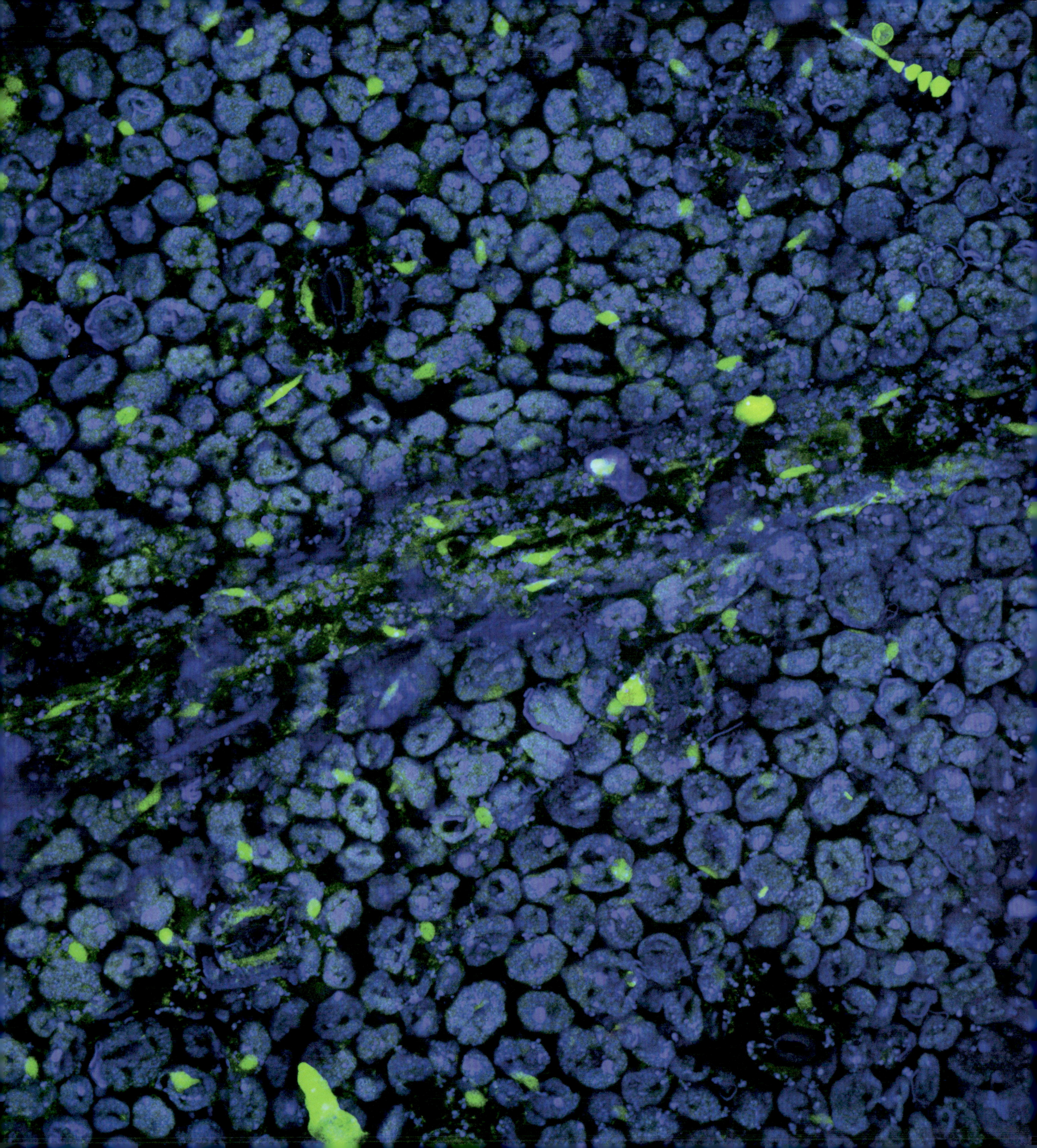

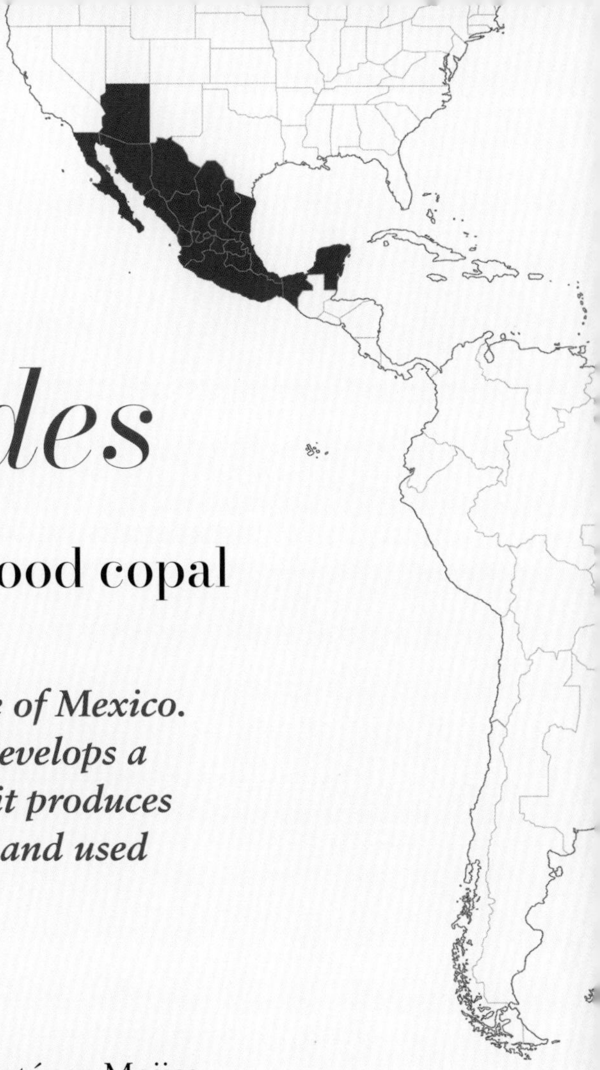

Burseraceae

Bursera fagaroides

Copal, Mexican frankincense, pom, torchwood copal

This deciduous small tree grows in the seasonally dry biome of Mexico. Its peeling, shaggy bark reveals a green trunk that often develops a distinctive swollen bole. Known as Mexican frankincense, it produces resin from which the sacred incense copal has been made and used ceremonially since ancient times.

According to a team of Mexican scientists led by Mayra Antúnez-Mojica, *Bursera fagaroides* has proven antitumour activity. In their overview, they affirm that "in general, lignans from *Bursera fagaroides* exhibited potent anti-cancer activity, although antitumour, anti-bacterial, anti-protozoal, anti-inflammatory, and anti-viral properties have also been described". The researchers also mention that the resin from several *Bursera* species has been used by Indigenous groups since ancient times in religious activities. There is another more recent study of *Bursera fagaroides*, headed by Nancy Mejía-Pérez and conducted in Mexico. It also recognises the plant's cytotoxic compounds, but notes that extracting these bioactive lignans located low in the bark endanger the life of this slow-growing and difficult-to-propagate tree. The goal of these scientists, therefore, "was to search for alternative sources of cytotoxic compounds in *Bursera fagaroides* prepared as leaves and in vitro callus cultures", which they were able to do successfully.

Copal is an incense used throughout Mesoamerica that is prepared from the resin (often a blood-red colour) from different trees of the Burseraceae family, including *Bursera fagaroides*, *Bursera bipinnata*, and *Bursera simaruba* (called Jiñocuago in Nicaragua).

Bursera fagaroides (copal); confocal image

50 μm

Microcosms – Sacred Plants of the Americas

"And here is the dawning and showing of the sun, moon, and stars. And Jaguar Quitze, Jaguar Night, Not Right Now, and Dark Jaguar were overjoyed when they saw the sun carrier.

It came up first.

It looked brilliant when it came up, since it was ahead of the sun.

After that they unwrapped their copal incense, which came from the east, and there was triumph in their hearts when they unwrapped it.

They gave their heartfelt thanks with three kinds at once:

Mixtam Copal is the name of the copal brought by Jaguar Quitze.

Cauiztan Copal, next, is the name of the copal brought by Jaguar Night.

Godly Copal, as the next one is called, was brought by Not Right Now.

They were crying sweetly as they shook their burning copal, the precious copal."

The three of them had their copal, and this is what they burned as they incensed the direction of the rising sun.

From the *Popol vuh*
(*The Mayan Book of the Dawn of Life*)
Translated by Dennis Tedlock

Bursera fagaroides (copal); Baja California Sur

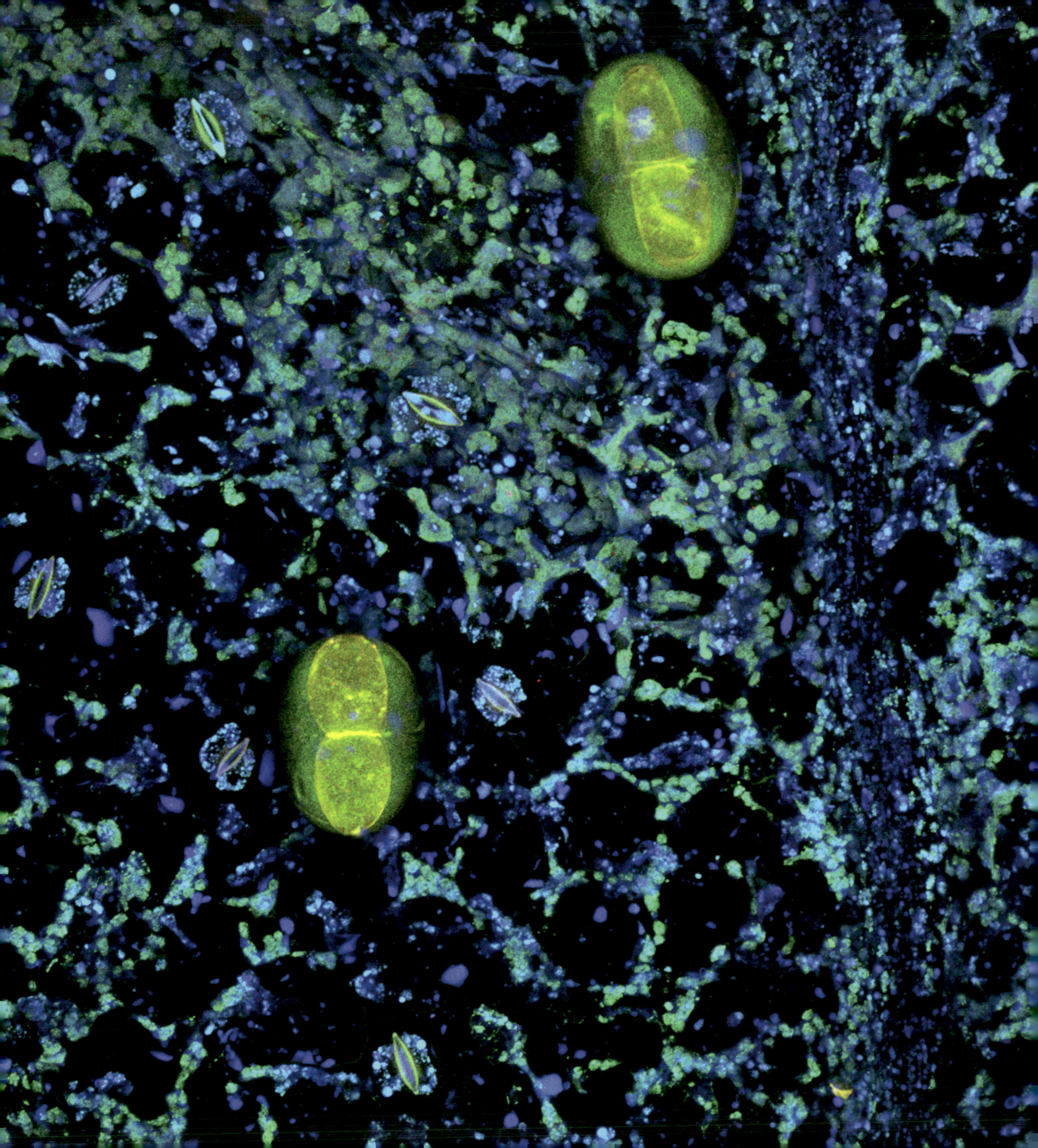

Asteraceae

Calea ternifolia

formerly *Calea zacatechichi*

Aztec dream grass, thlé-pelakano, zacatechichi

This herbaceous plant has small oval leaves (with a bitter taste) that are coarsely toothed on the edges, and prefers the climate of the highlands of central Mexico. Chontal healers call the plant thlé-pelakano, or "leaf of god". Contemporary researchers call it an "oneirogen", a plant that produces lucid dreams.

Calea zacatechichi should now be called *Calea ternifolia*, an earlier name established in 1820 that taxonomically preceded that of *Calea zacatechichi*. Even so, the Náhuatl etymology of *zacatechichi* is worth remembering since it describes one of the plant's most memorable qualities as "*bitter* grass". The extremely bitter taste of infusions prepared from its leaves has not kept *Calea* from being used as a folk medicine for the treatment of headaches, diabetes, and gastrointestinal disorders. It grows primarily in the mountainous regions of Mexico, and is used by the Chontal Indians of Oaxaca, according to Christian Rätsch, to produce "dreamlike states in which they can hear the voices of the gods and spirits, recognise the causes of illnesses, look into the future, and locate lost or stolen objects".

Lilian Mayagoitia, José-Luis Díaz, and Carlos M. Contreras explain the need for their 1986 scientific investigation of *Calea*, saying "the use of plant preparations in order to produce or to enhance dreams of a divinatory nature constitutes an ethnopharmacological category that can be called *oneiromancy*, and which justifies rigorous neuropharmacological research". They found that in human volunteers, *Calea* extracts "increased the superficial stages of sleep and the number of spontaneous awakenings", and, based on subjective reports of dreams, "an increase in hypnagogic imagery".

Díaz created a psychopharmacological classification system for sacred plants. He placed *Calea* in a category he calls: "cognodysleptics – marijuana (tetrahydrocannabinol)

Calea ternifolia (zacatechichi); confocal image

50 μm

and other terpene-containing plants induce changes in thought, imagination, and affective functions, and are used in short-term divination or oneiromancy".

Although both Rätsch and Jonathan Ott, based on their own self-experiments, are dubious about *Calea* as a full-blown entheogen, a small army of determined psychonauts, writing over a period of some twenty years with bitter honesty about the foul taste of this plant ingested as an infusion or smoked, describe powerful lucid dreaming in the pages of the *Erowid Experience Vaults*. Surely, the work of these valiant citizen-scientists counts for something!

A study by L. Martínez-Mota et al. demonstrated for the first time "that *Calea zacatechichi* produces strong and specific anxiolytic-like effects [and also] favours some sleep stages related with cognitive processes, such as memory consolidation". They believe that, in the future, this plant's antidepressant qualities "could improve some aspects of mental health".

Research from 2022 conducted by R. Mata and his colleagues highlights the sesquiterpene lactones as the most important metabolites of the Mexican "Dream Herb". They conclude that pharmacological studies have proven "the antinociceptive, anti-inflammatory, spasmolytic, antiprotozoal, antidepressive, antidiarrheic, anxiolytic, and the antidiabetic properties of different preparations of the plant", many of which coincide with the traditional uses of *Calea*.

In an article on the neuropsychopharmacological induction of lucid dreams, a Brazilian team of researchers led by Abel A. Oldoni mentions that previous studies have established that "*Calea* induced a discrete increase in all sensorial perceptions, discontinuity in thoughts, a rapid flux of ideas with difficulty in their retrieval, and statistically significant slowness of reaction time, which might have induced a light hypnotic state", and also that "one of the mechanisms of action of *Calea* is through its sesquiterpene lactones". New studies are demonstrating the therapeutic potential of sesquiterpenes for the treatment of Alzheimer's disease on account of their Acetylcholinesterase Inhibitory (AChEI) properties. The researchers from Brazil speculate that it may be these very AChEI properties that enable *Calea* to induce lucid dreaming.

above: *Calea ternifolia* (zacatechichi); plant
opposite: *Calea ternifolia* (zacatechichi); confocal image

50 μm

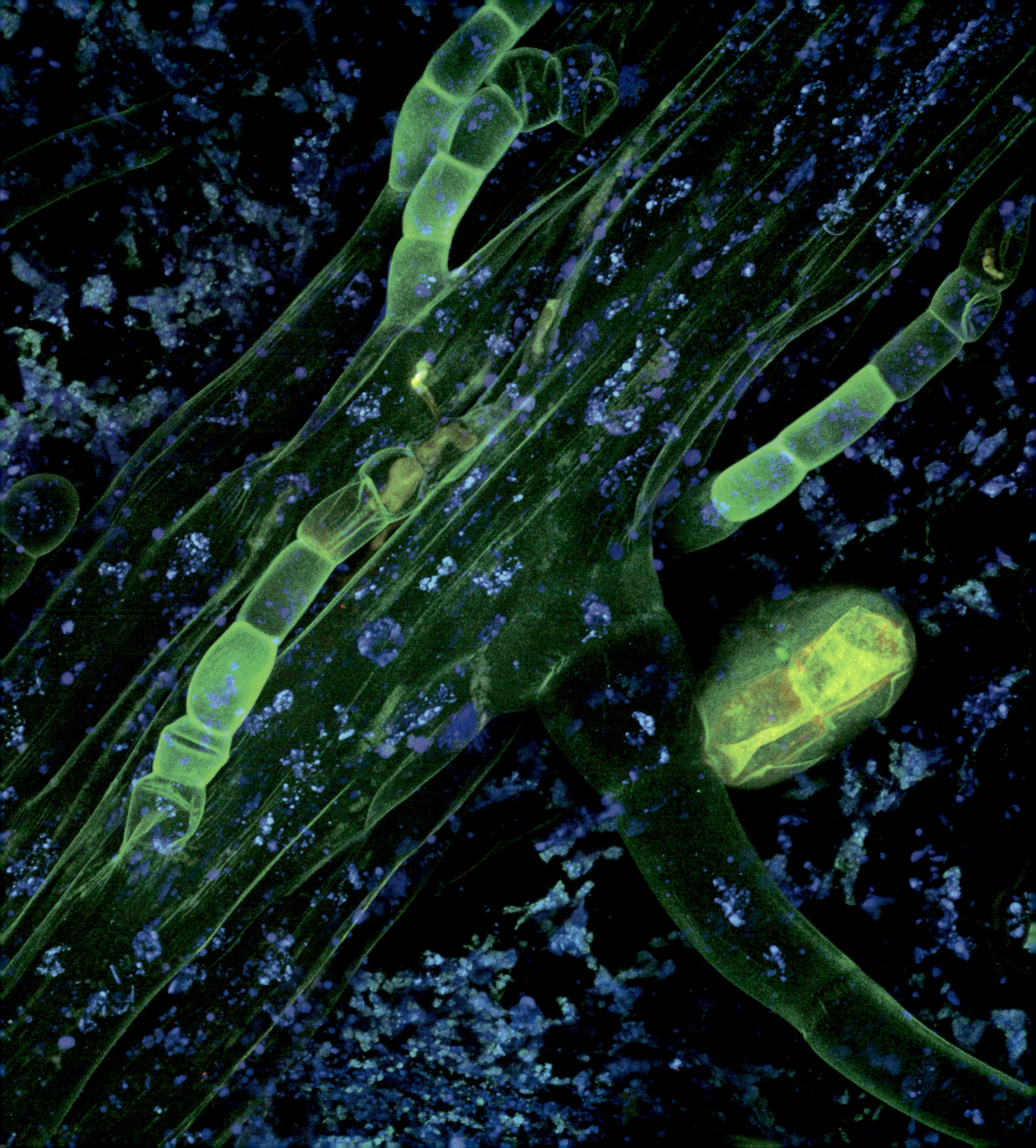

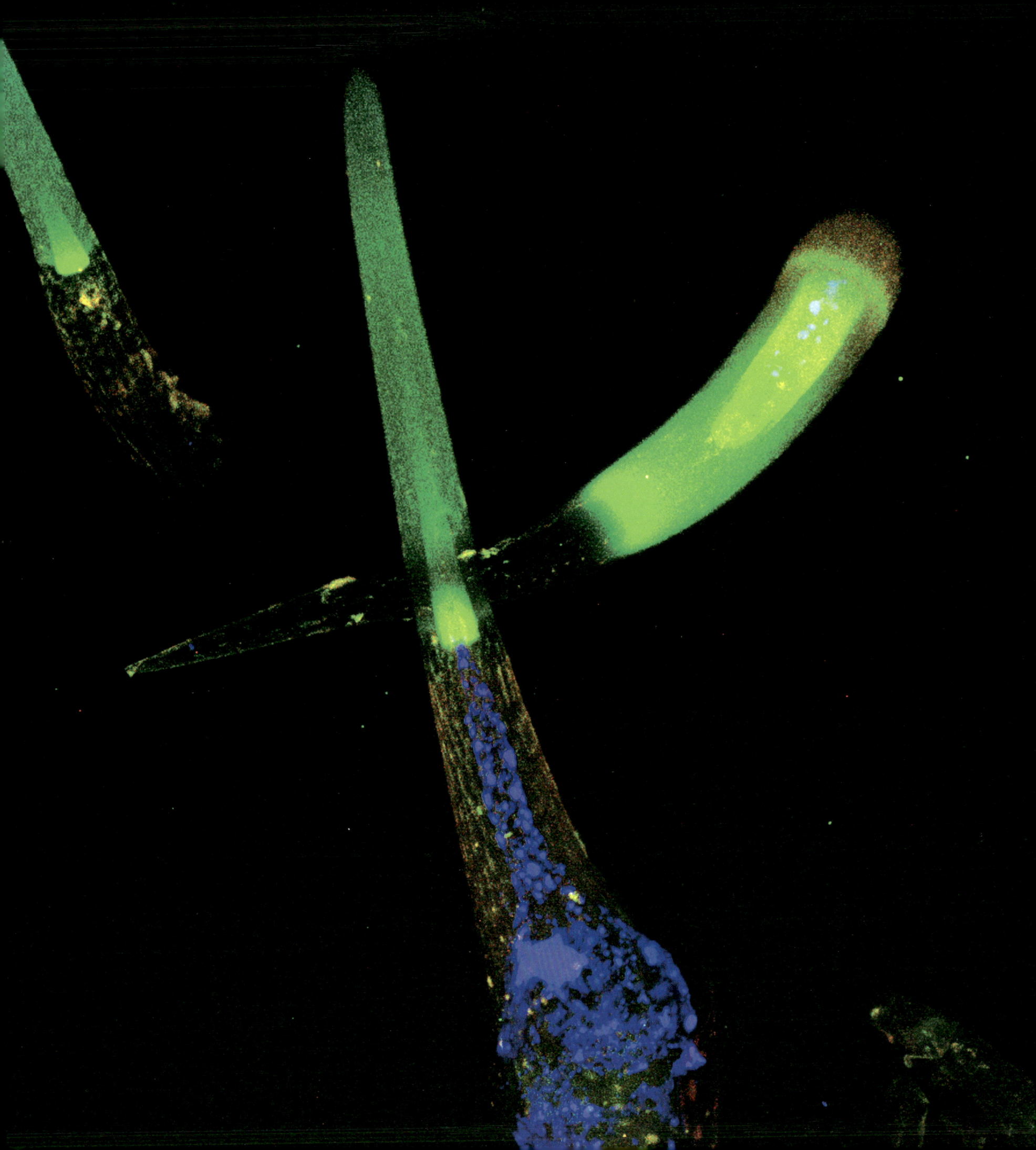

Cannabaceae
Cannabis spp
Marijuana, monte, pot, weed

The authors of Plants of the Gods described this polymorphic plant as a loosely-branched annual herb with membranaceous leaves that are digitate with 7 to 9 linear-lanceolate serrated segments and flowers that are borne in terminal branches. Then there was a small flame and their words went up in smoke!

Schultes and Hofmann (with Rätsch), the authors of *Plants of the Gods*, write of a relationship between *Cannabis* and humanity that has "existed now probably for ten thousand years – since the discovery of agriculture in the Old World". They document ancient uses of this multi-purpose plant in India, China and elsewhere.

Cannabis was taken to many regions throughout the world, eventually arriving in the Americas: "rarely is an introduced foreign plant adopted and used in Indigenous ceremonies, but it seems that the Cora of Mexico and the Cuna of Panama have taken up the ritual smoking of cannabis, notwithstanding the fact that, in both areas, it was brought in by the early Europeans".

A CNN interview by Dr. Sanjay Gupta with Dr. Staci Gruber (also known as the "Pot Doc"), an Associate Professor at Harvard Medical School and Director of the Cognitive and Clinical Neuroimaging Core, as well as the Marijuana Investigation for Neuroscientific Discovery program, emphasises some remarkable discoveries about *Cannabis* and its medical benefits. According to Dr. Gruber: "it's really helpful to take a step back and to acknowledge that the plant is this miraculously complex structure with over 500 compounds, and it does appear in pre-clinical work and some of the basic experiments that have been done that certain cannibinoids, in combination with other cannibinoids and other compounds, appear to allow things like tumour progression to either slow, halt, or even be reversed. That's extraordinary!" About the current efforts

Cannabis sativa (marijuana); confocal image

50 μm

to move *Cannabis* from Schedule I of the Controlled Substance Act of 1970 to the less restrictive category of Schedule III which would recognise the medicinal value of *Cannabis* and facilitate new scientific research, Gruber says that it is important to note that "reschedule is not deschedule". Schedule III would mean changes in security requirements, monitoring, and surveillance for scientists. Such a shift could result in more studies that reveal medicinal benefits as astonishing as the treatments that were derived from *Cannabis* to alleviate paediatric onset intractable seizure disorders in children. As Gruber asserts: "there's a lot of work to be done for folks who are experiencing mild cognitive impairment and different neurodegenerative disorders". There are further studies listed in the Bibliography which detail the impact of medical cannabis treatment on chronic pain, as well as, for example, an overview of *Cannabis sativa* as a neutraceutical.

above: *Cannabis sativa* (marijuana); closeup of two kinds of trichomes on a leaf
opposite: *Cannabis sativa* (marijuana); confocal image

100 μm

overleaf left and right: *Cannabis sativa* (marijuana); plant

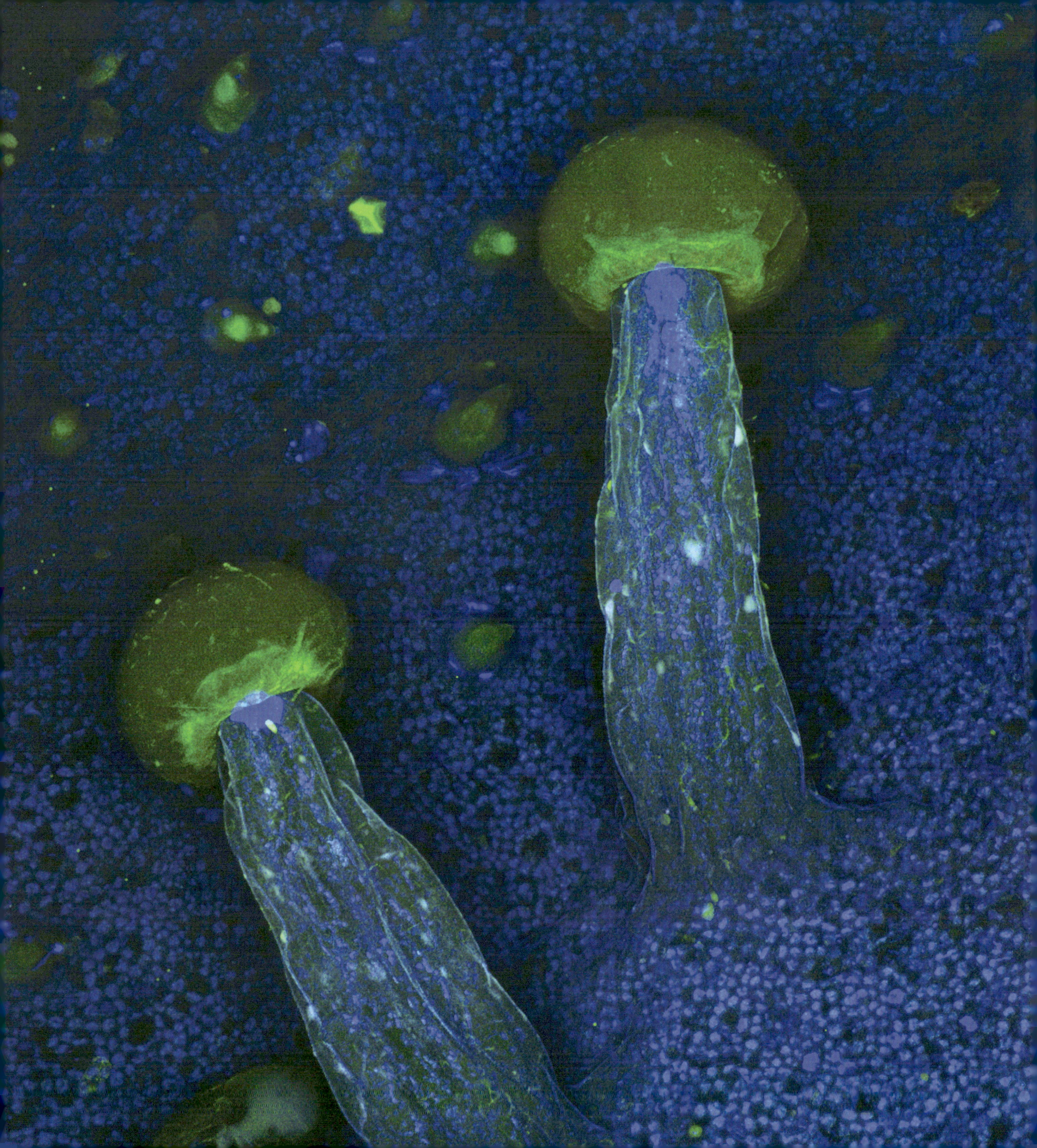

Malvaceae
Ceiba pentandra

Ceiba, kapok, lupuna, samaúma, silk cotton tree,
wari mahi, yax che el cab

A giant of the rainforests, this fast-growing deciduous tropical tree can reach astonishing heights, and produces an abundance of seeds attached to silky, water-resistant fibres. It has many uses for humans, from its lightweight wood and fibres to its seed oils. The Maya revered this sacred tree as "The Tree of Life", believing it connected the heavens, earth, and the underworld.

One January, my wife Esthela Calderón and I secretly rushed ceiba (as well as cacao and copal) leaves wrapped in damp paper towels and sealed in plastic bags across borders from our ancestral farm in Pueblo Redondo, Nicaragua all the way back to wintry Canton, NY and the confocal microscope at St. Lawrence University where Jill Pflugheber was poised for action. We were all ecstatic with the botanical forms (particularly the stomata and trichomes), as well as the colours, juxtapositions, and evident raw power of this towering tree in no way diminished by the microscope and the hum of electrical current; this sacred emblem of protection, Axis Mundi, joiner of earth and sky, revealed at last in these images.

No one has written a more beautiful ethnobotanical portrait of a ceiba than Nicaraguan poet Pablo Antonio Cuadra (1912-2002), my literary mentor for decades. The poem, which I translated with Greg Simon, is one of my absolute favourites by Cuadra, and comes from the extraordinary book *Seven Trees Against the Dying Light*. Overleaf is a fragment from "The Ceiba Tree".

Writing about Upper Amazonian shamanism in Peru, Françoise Barbira Freedman says that tobacco is offered as a propitiating food to the mother-spirits of certain trees, particularly the *lupuna* tree (*Ceiba* spp.): "the *lupuna* sap is indeed known to be poisonous as well as psychoactive". I have written more about these properties of the ceiba in a short essay that appears in *The Mind of Plants*.

Ceiba pentandra (ceiba); Finca Santa Ana, Pueblo Redondo, Telica, Nicaragua

above: *Ceiba pentandra* (ceiba); seed pods, Florida
opposite: *Ceiba pentandra* (ceiba); a medicine tree from the Peruvian Amazon known as lupuna blanca
Look for the purging jaguar in the roots near the ascending *Alicia macrodisca* vine
overleaf left and right: *Ceiba pentandra* (ceiba); confocal image

50 μm

In a book on the shamanic practices of the Yanomami, Bruce Albert and William Milliken affirm that the Indigenous healers use the "images" of the largest trees of the Amazon rainforest, such as the ceiba, to scare off evil spirits that cause disease. What if these powerful confocal images could serve the same purpose?

The geographical parameters of this book are limited to the American continent, yet some of the plants that are represented have a global presence, growing in similar ecological niches around the world. Such is the case of *Ceiba pentandra*, and, try as one might to focus on the sacred qualities of this tree in different Amerindian contexts, there is substantial research currently being done on the industrial uses of the kapok, silk cotton, and java cotton tree in Asia and Africa. These include the production of biofuels, biogas, biocatalysts, biocomposites, and cellulosic textiles of a natural origin in the interest of reducing environmental pollution that is the result of synthetic industrial waste. This is the contemporary ethnobotanical fate of the towering giant whose branches seem to touch both sun and moon, and whose roots extend into the Mayan underworld of Xibalba where the Lords of Sickness and Death engage in mortal combat with the Hero Twins of the *Popol vuh*.

Ceiba pentandra

"This tree was born in the centre of the world.

From its highest branches you see what your heart longs for.

This is the tree that lovingly cradles your childhood on its lap.

With the light, silky cotton of its fruit, your people made the pillows on which they rest and shape their dreams.

Climbing this tree, the serpent becomes bird and the word, song.

This is the Mother Ceiba in whose swelling trunk your people honoured birth and fertility.

From a single piece of its white, easily carved wood, they built a vessel that is their cradle when their journey begins and their coffin when they reach their port.

From this tree, humanity learned mercy and architecture, order, and how to give with grace."

Pablo Antonio Cuadra

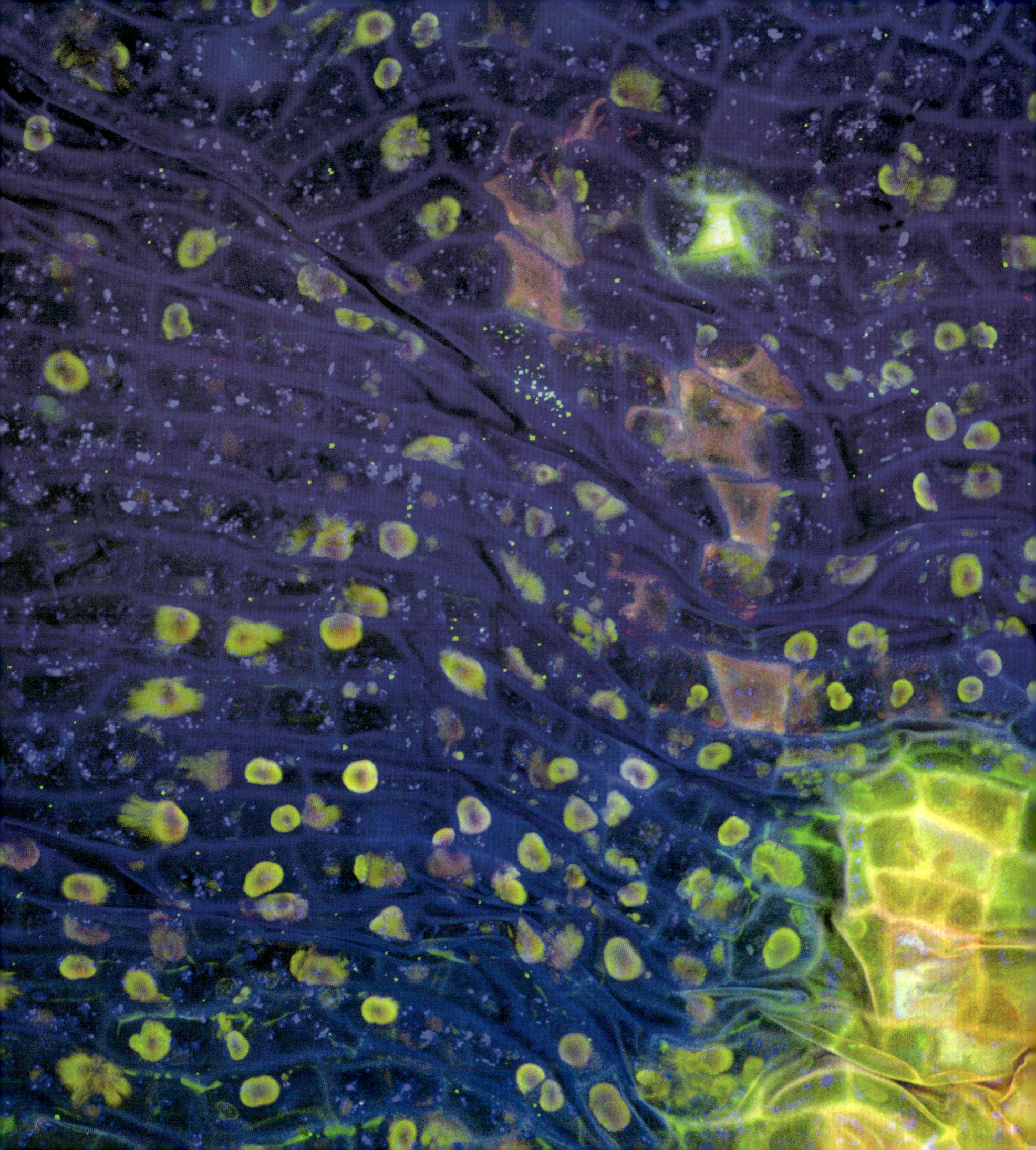

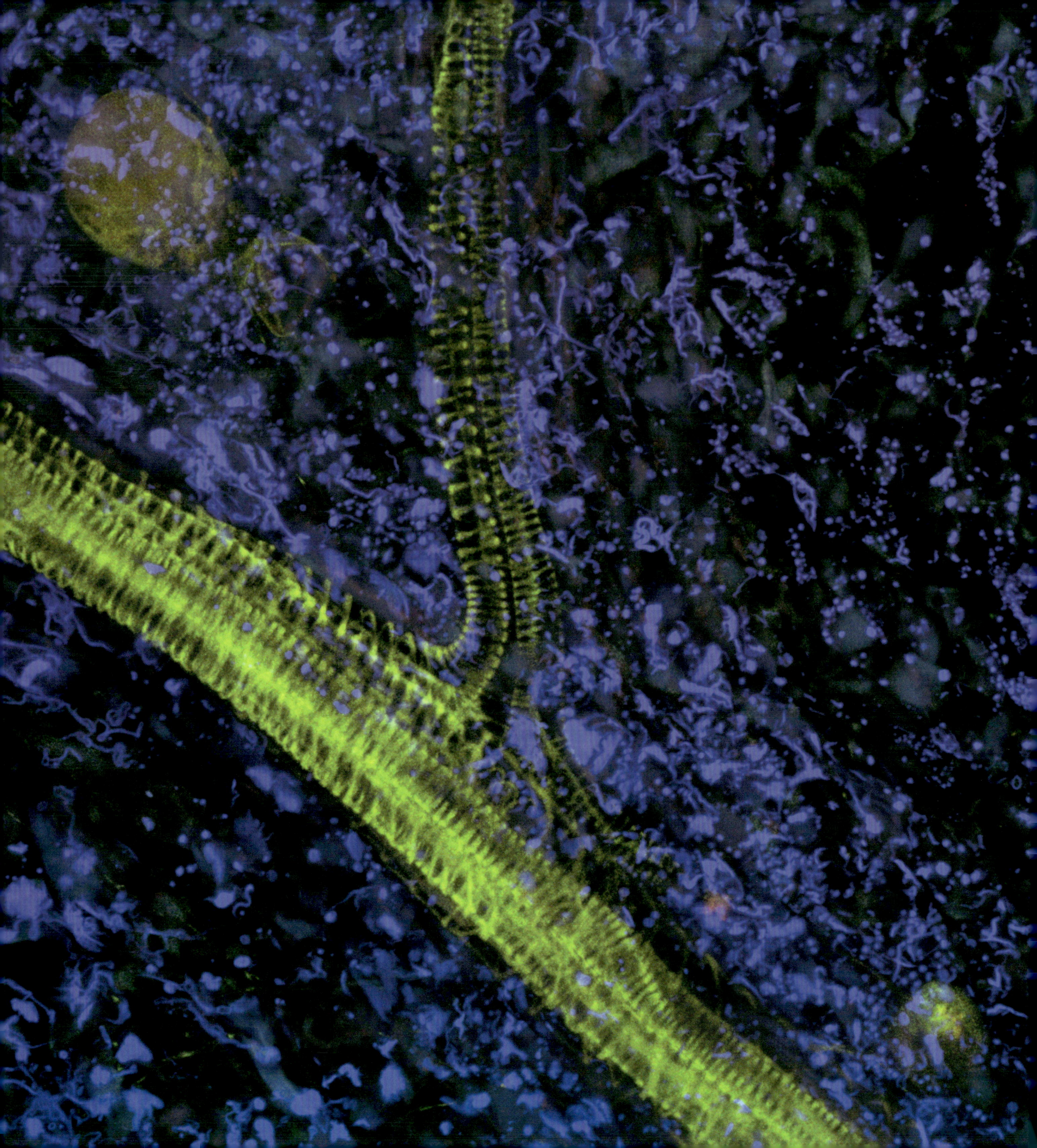

Solanaceae

Cestrum parqui

Chilean cestrum, Chilean flowering jessamine, hediondilla, palqui

This semi-evergreen fast-growing shrub has thin, upright, woody stems and oval leaves with a distinctly unpleasant odour when touched, earning it the common name hediondilla ("little stinker") in Spanish. The Mapuche-Huilliche in southern Chile use this plant, known as palqui, in purification rituals to banish evil spirits and enemy shamans.

Cestrum parqui, commonly known as *palqui*, is a flowering bush native to central Chile whose fetid leaves have been used medicinally by the Mapuche for the treatment of wounds, rashes, allergies, inflammations, and fevers. Mössbach mentions the Chilean saying: "wherever the devil has planted a nettle, God has planted a *palqui*". In a recent phytochemical overview of this plant, Bahgat et al. describe how *Cestrum* L. species are responsible for cytotoxic, spermicidal, anti-microbial and pesticidal activities. Huanquilef et al. demonstrate in their study how *Cestrum parqui* can control insects that have a negative impact on the Chilean forestry sector. *Cestrum parqui*, affirm the researchers, could be an alternative treatment, replacing deleterious methyl bromide fumigation that has ozone-depleting effects on the environment.

But there are additional characteristics of this plant that give it a special importance to the Mapuche culture. In an anthropological study of the *ruka*, the Mapuche dwelling in the forested mountains of southern Chile, Juan Carlos Skewes describes how homes constitute part of a living landscape, a way for their inhabitants to become integrated with the environment and thereby protect both the forest and themselves. Skewes discusses how the trees and shrubs become allies of the Indigenous residents in their everyday lives "as a source of wisdom or health in the case of the *pellín* (*Nothofagus*

Cestrum parqui (palqui); confocal image

50 μm

left: *Cestrum parqui* (palqui); in bloom
above: *Cestrum parqui* (palqui); also known as hediondilla ("the little stinker")

obliqua), whose physical presence is advisable for the sick, or as a *contra* (antidote) for evil spells in the case of *palqui* (*Cestrum parqui*), a shrub known for its toxicity".

There is a certain secrecy that seems to envelop this plant, though this is certainly understandable given the prying eyes of non-Mapuche outsiders. Plowman mentions *Cestrum parqui* in an article on another powerful member of the Solanaceae family *Latua pubiflora*: the Mapuche-Huilliche of southern Chile hold purification and healing ceremonies that can include whipping the patient with the branches of the foul-smelling *palqui* (known in Spanish as *hediondilla*) to banish the evil spirits and enemy shamans who have caused the illness. Rätsch affirms that he experienced the psychoactive properties of the smoked leaves of *palqui*, comparing the atropine-like effect to that of the solanaceous *Brugmansia*. Rätsch also cites sources that name *Cestrum parqui*, along with *Latua pubiflora* and other plants, as a principal ingredient in a psychoactive incense used ritually by the Mapuche.

According to a fascinating book edited by Iván Pérez Muñoz, the most sacred place for the Mapuche-Lafkenche is Isla Mocha, an island in the Arauco Province of Chile, 40 kilometres off the coast of Tirúa. The name of this island, known by the Indigenous population as Amucha or

Amuchura, comes from the words in the Mapuche language Mapudungun *Am* ("soul") and *Uchran* ("to be resurrected"). This is the stopping place on the way to Wenumapu, the Mapuche Paradise. As the story goes, it is easy enough to live and much more difficult to die. Nomtufe ferries the souls with their little flames in the immensity of the night from the continent to the island in a dugout canoe made of a single piece of the Chilean Laurel (*Laurelia sempervirans*) that is called *triwe* by the Mapuche. Four venerable women transform themselves into whales (*meli yene*) at the end of the day, and follow the embarkation as they carry on their backs an irascible woman with long white hair known as the Trempülkahue, the Supreme Judge, the Interrogator of Souls. The whales make the trip increasingly difficult with the currents and whirlpools they create, demanding payment for the transportation provided in the form of a necklace of precious stones carried by the souls. As the turbulent journey continues, each of the soul's two eyes make a small splash into the water as they sink beneath dark waves and calm the Trempülkahue, who then decides which of the souls are worthy of continuing to Paradise as *kimche* – people who were good, just, and hardworking in life, who had a sense of lineage and of belonging to a place, and who possessed spiritual fortitude. An archaeobotanical study conducted by Chilean researcher Carolina Godoy-Aguirre at a site on the northeastern part of Isla Mocha known as El Vergel Complex examined microscopic plant residue on ceramic shards and positively identified five species, including *Zea mays* (corn) for fermented drinks (in approximately 1300 CE) as well as *Cestrum parqui* used for medicinal or for as yet unknown ritual purposes. In 1685, in another instance of colonial repression, Spanish authorities forced the 500 Indigenous residents of the island to abandon the sacred ground of their ceremonial centre and sent them to live at a missionary settlement in Concepción. Isla Mocha may have been uninhabited for the next 160 years, though some believe that a small group of Mapuche permanently remained in the part of the island that is now the Nature Reserve.

In a paper published in 2002, a group of scientists from Texas and Bolivia headed by Robert R. Luedtke, who hoped "to identify novel compounds that modulate dopamine receptor activity", discovered that an aqueous extract of *Cestrum parqui* "was found to contain a stable component that appears to be an agonist at D1-like dopamine receptors" and also had "intrinsic activity at D2-like dopamine receptors". This makes *Cestrum parqui*, of the more than fifty plants tested for the study, the strongest candidate for use in the treatment of illnesses such as Parkinson's disease, Tourette's syndrome, schizophrenia, and cocaine addiction. A group of Italian researchers led by Maria Chiara Di Meo published an overview of *Cestrum parqui* in 2024 in which the scientists summarise the studies of "antimicrobial, anticancer, insecticidal, antifeedant, molluscicidal, and herbicidal properties" of the leaves. They conclude that their research "justifies the significant interest in this plant, with possible and concrete commercial application".

Solanaceae
Datura innoxia
Chamico, downy thorn apple, toloache

Datura blooms at night, and has upright, fragrant flowers. Its fruit is covered with sharp spines. Rich in tropane alkaloids, it was used by many Mesoamerican Indigenous peoples to produce ceremonial visionary states. Contemporary scientists have documented Datura's wide-ranging medical efficacy.

According to Peter T. Furst, "*Datura toloache* from the Nahuatl *toloatzin*, in Mexico and also in Indian California, was, and in many places still is, the ritual intoxicant of choice among native peoples of the Southwest and northwestern Mexico, including the Tepehuan". Also called downy thorn apple, this plant was used by the Aztecs (Mexica) to reduce fever, by the Tarahumara (Rarámuri) to fortify fermented drinks, and by the Yaqui (Yoeme) to produce a visionary state.

This sacred plant is associated with numerous Indigenous myths. For example, the authors of *Plants of the Gods*, Schultes and Hofmann, recount the Zuñi Indian story about the divine origin of *Datura* in which a brother and sister knew too much about ghosts and the hidden things of the world and, consequently, offended the Divine Ones, who banished them forever. The *Datura* flowers appeared where the two descended into the earth. The blooms were exactly the same as the ones with which the brother and sister would adorn themselves on each side of their heads when they used to visit the outer world.

Now that it is possible to include a more ample selection of confocal microscope images, it is clear that some species really seem more "photogenic" than others. *Datura* really is a star, perhaps attributable in part to the fact that it was growing in a close friend's garden in upstate New York and was not one of the species that had to be transported from afar. A fresher specimen would not be possible! Especially notable

Datura innoxiae (toloache); confocal image

50 μm

are the grains of pink-tinted pollen, and the striated textures of the vascular tissue in which the stomata are embedded. When botanical structures cede to the purely abstract dimension of *Microcosmic Phytoformalism* as they do in many of the *Datura* images here, these perfectly natural shapes that we have documented are every bit as interesting to contemplate aesthetically as the visual expression produced by professional artists.

Guillermo Benítez, from the Department of Botany, Faculty of Pharmacy, at the University of Granada led a team of Spanish and Mexican scientists in a study of the genus *Datura* from what the researchers call "an ethnobotanical perspective at the interface of medical and illicit uses". The article highlights "historical knowledge from post-colonial American Codices [e.g. the *Badianus manuscript* and *Florentine Codex*] and medieval texts", and also discusses "*Datura*'s current social emergency as a drug of recreation and leisure, as well as its link to crimes of sexual abuse" due to the "incapacitating" and "amnesia-producing properties" of its alkaloids, something that the authors believe is of "maximum relevance in the field of forensic botany and toxicology". In Spain, *Datura* is considered an emerging drug linked to both recreational use and an increasing number of crimes against sexual freedom, specifically as "a substance used in cases of chemical submission for sexual purposes". The scientists believe that further research regarding "the ethnobotanical and ethnopharmacological knowledge about these plants" could produce "an advance in the medical research and in the standardisation of safer protocols" as well as a reduction in "severe cases of intoxication".

Researchers led by Meenakshi Sharma from the Department of Chemistry at Ranchi University in India conducted a review in 2021 "to summarise the phytochemical composition and pharmacological and toxicological aspects of the plant *Datura*". The scientists cite numerous studies documenting the medicinal efficacy of *Datura* as possessing "antimicrobial, antidiabetic, anti-asthmatic, anti-inflammatory, antioxidant, analgesic, insecticidal, cytotoxic, wound healing, and neurological activities". Rich in the tropane alkaloids scopolamine, hyoscyamine, and atropine, Datura is also used for "mystic and religious purposes", and as a means to achieve a "hallucinogenic experience" that can produce toxic, extremely harmful adverse effects such as "fever, dry skin, dry mouth, headache, hallucinations, convulsions, rapid and weak pulse, acute confusion, delirium, tachycardia, coma, and death". These same powerful alkaloids, as muscarinic antagonists, can also be used to cure Parkinson's disease, and as a therapy for asthma symptoms through the plant's bronchodilating effects when its leaves are smoked. In their conclusion, the authors highlight *Datura*'s "use in ayurvedic medicine for the treatment of wounds, inflammation, bruises and swellings, sciatica, ulcers, rheumatism, asthma, bronchitis, and body ache", and reiterate how the plant's "toxic effects generally conceal its medicinal effects".

top: *Datura innoxiae* (toloache); seed pod
above: *Datura innoxiae* (toloache) in bloom
opposite: *Datura innoxiae* (toloache); confocal image

50 μm

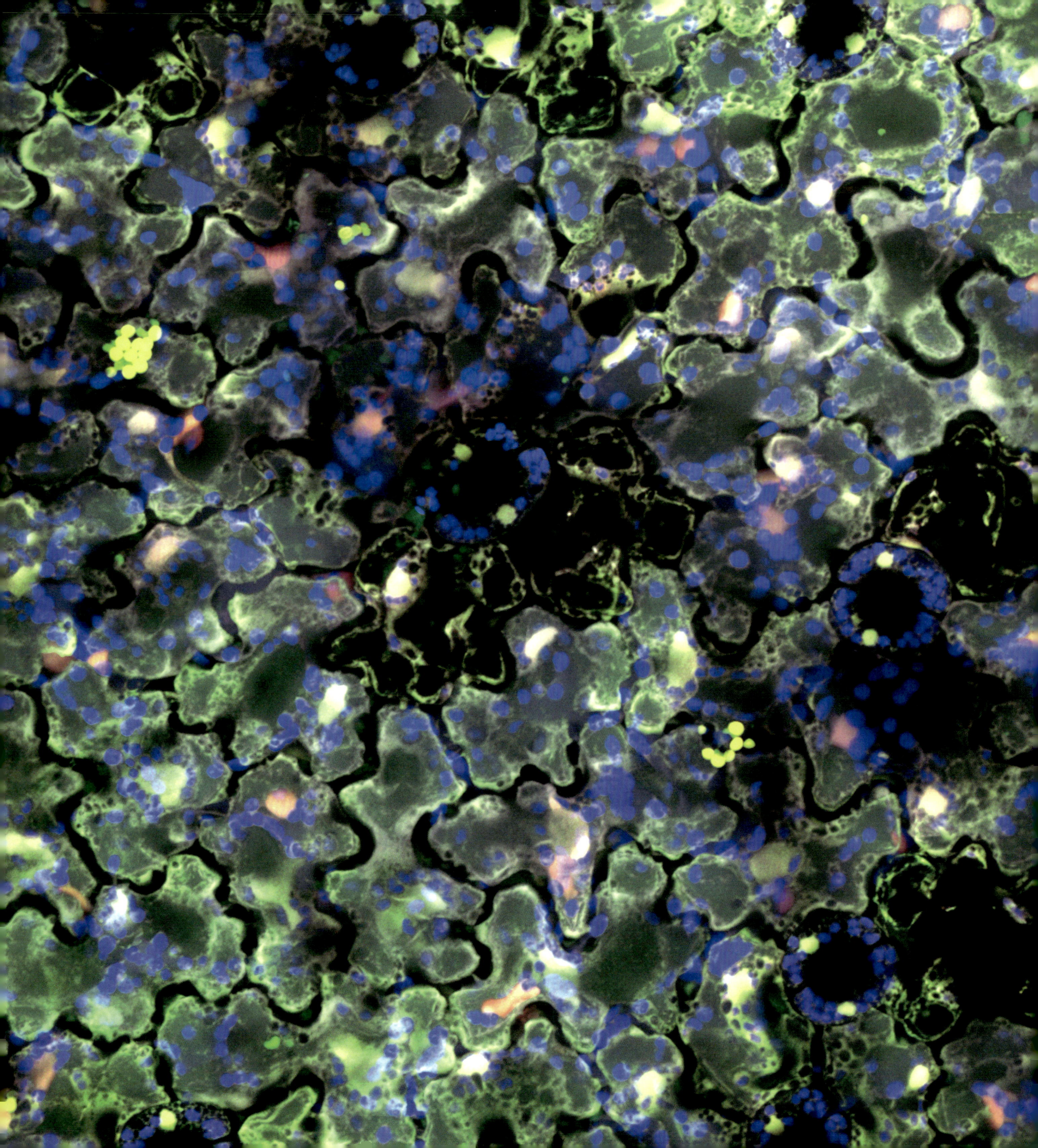

Collumelliaceae
Desfontainia spinosa

Borrachero de páramo, chapico, michai blanco, taique, trau-trau

This subtropical evergreen bush has thick, thorny, dark green foliage and funnel-shaped orange-red flowers. Its leaves strongly resemble those of English holly. The legendary ethnobotanist Richard Evans Schultes studied this plant during his research trips to Colombia and became fascinated by its aura of mystery.

Richard Evans Schultes, in his pioneering article from 1977 "*Desfontainia*: A New Andean Hallucinogen", describes collecting *Desfontainia spinosa* twice in Colombia's Sibundoy Valley, first in 1942 and then in 1953. The Kamsá and Ingano shamans that Schultes consulted called the plant *borrachero de páramo* and told him that they would drink a tea from its leaves when they "want to dream", and also "to see visions and diagnose illness". Generally, however, Schultes found that local medicine men were reluctant to discuss this plant's use, writing "this reluctance in itself is an indication possibly that its employment is held more in secret because of a very special place that the plant holds in magico-medical practice". The plant, also known as Chilean Holly due to the shape of its leaves (which are used by the Mapuche in southern Chile to create a yellow dye for wool used in the making of traditional clothing), still retains many of its mysteries. A team of researchers from Chile and Spain led by Emir Valencia studied the antifeedant activity of *Desfontainia spinosa* and found that it effectively had a deterring effect on the insect *Leptinotarsa decemlineata*. However, despite the extensive phytochemical study conducted, the scientists concluded that "none of the isolated products provide a basis for the reputed hallucinogenic activity" of this plant.

Desfontainia spinosa (taique); confocal image

50 μm

"Many Indigenous groups around the world – the Indians of the Amazonian regions, for example – are literally masters of their ambient vegetation as a result of inherited knowledge. This knowledge – of great potential value to humanity as a whole – is unfortunately doomed to extinction with the rapid acculturation and westernisation in many parts of the globe where Indigenous peoples can still live peacefully without disruption, from road-building, airstrips, missionary pressure, warfare, tourism, industrial penetration, dam-building, local greed on the part of settlers, or various efforts to "civilise" the natives."

~ Richard Evans Schultes

opposite: Richard Evans Schultes on the summit of Cerro Campana, Guaviare, Colombia, June 5, 1943
right and below: *Desfontainia spinosa* (taique); flowering
overleaf left and right: *Desfontainia spinosa* (taique); confocal image

50 μm

Desfontainia spinosa

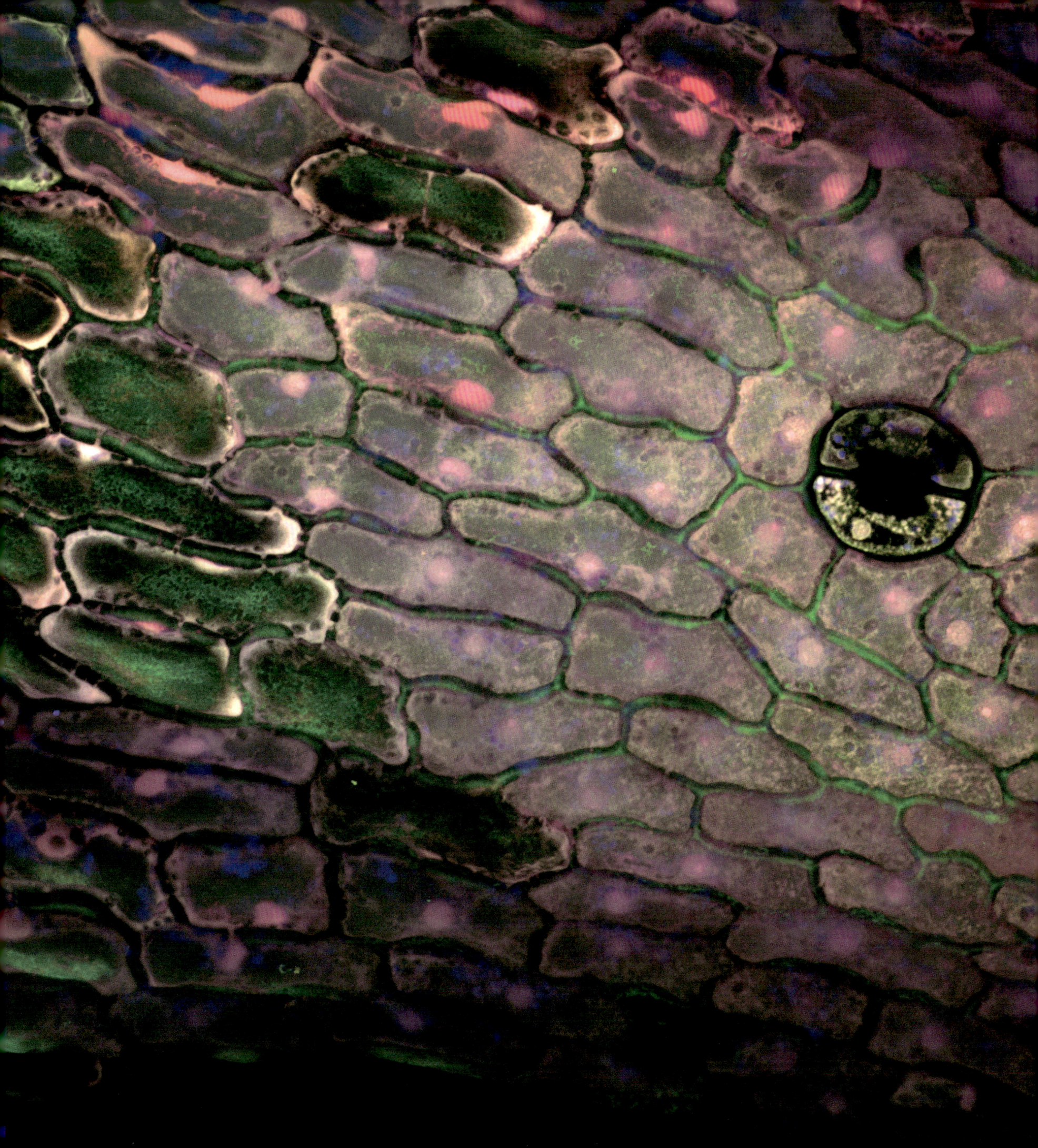

Acanthaceae
Dianthera pectoralis

Carpenter bush, chambá, chapantye, mashi hiri

A helophyte that grows in the seasonally dry tropical biome, it has leaves that contain coumarins (which give them a pleasant fragrance), and abundant, small, light-purple orchid-like flowers. The Yanomami add the leaves to their preparation of the shamanic Virola snuff Yãkoana, which opens the path to the spirit worlds.

The Yanomami use the shade-dried leaves of *Dianthera pectoralis* as an additive to psychoactive *Virola* snuffs. This unassuming plant that grows throughout the Caribbean and Central and South America with a wide variety of common names reveals a microcosmic world that is shockingly complex and beautiful. It has a plethora of medicinal uses that include alleviating prostate problems, coughs and colds, skin rashes, diabetes, menstrual pains, menopause, epilepsy, and respiratory tract disorders. Due to the presence of coumarin, the leaves of *Dianthera pectoralis* have a sweet smell. Recent phytochemical research conducted by Luzia Kalyne Almeida Moreira Leal et al found that the plant "has therapeutic potential for the treatment of inflammatory diseases such as asthma".

A team of Brazilian researchers led by Thays Lima Fama Guimarães studied *Dianthera pectoralis* for the journal *Food Chemistry Advances* in 2023 and concluded that this plant, known as chambá in Brazil, "demonstrated great potential, presenting a composition rich in phenolics, especially umbelliferone, presenting a higher antioxidant activity and antimicrobial action, which was visualised by its minimum inhibitory and bactericidal concentrations on important foodborne bacteria".

Dianthera pectoralis (carpenter bush); confocal image

50 μm

above: *Dianthera pectoralis* (carpenter bush); flowers
opposite: *Dianthera pectoralis* (carpenter bush); confocal image

50 μm

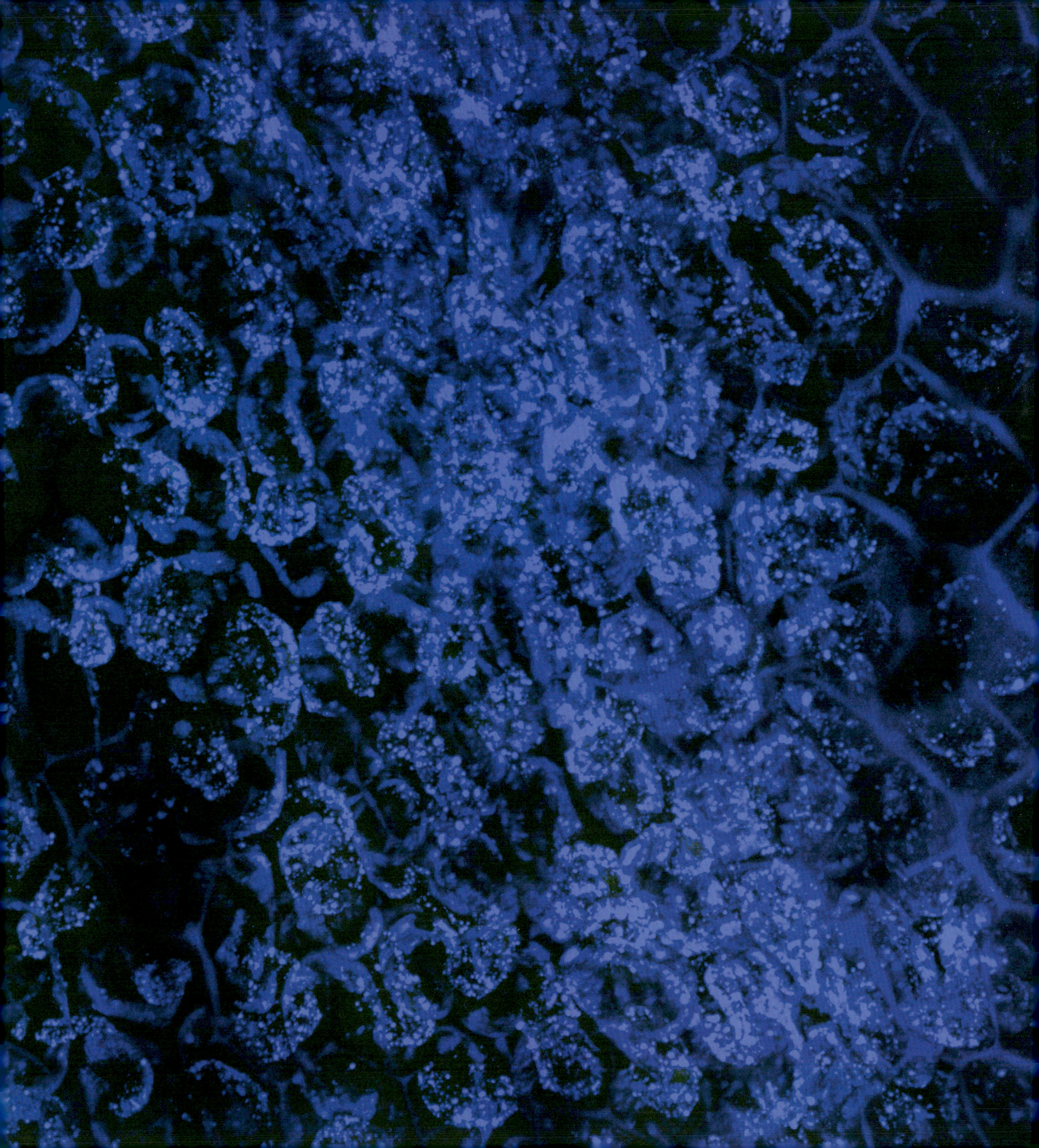

Winteraceae
Drimys andina

Canelo enano, canelo andino, dwarf winter's bark

Drimys andina is a closely-related dwarf species of Drimys winteri (also known as canelo and foye), the most culturally-significant tree of the Mapuche in the temperate biome of southern Chile and Argentina. It has grey bark, long leathery leaves, and white star-shaped flowers. Recent scientific studies have demonstrated that Drimys has antioxidant, anti-inflammatory, antimicrobial, antiparasitic, and antiviral properties, and effectively reduces triglycerides and total cholesterol.

One of the last plants we were able to image, *Drimys andina* (with thanks to Sacred Succulents in California), enabled us to extend the geographical representation of sacred plants in our project much further south into the immense forests that are the Mapuche ancestral lands on both sides of the cordillera of the Andes in Chile and Argentina. A synonym for *Drimys andina* is *Drimys winteri* var. *andina*. The shrub is a smaller version of *Drimys winteri*, which is highly revered by the Mapuche and known as canelo and foye. According to Ana Mariella Bacigalupo, "*foye* trees are sacred trees of life that connect the natural, human, and spirit world, and allow Mapuche shamans, or *machi*, to participate in the forces that permeate the cosmos. They are symbols of *machi* medicine, and *machi* use the bitter leaves and bark to exorcise evil spirits, as an anti-bacterial for treating wounds, and to treat colds, rheumatism, stomach infections, and ringworm".

Over the years, as a literary critic, translator, and editor, I have appreciated and promoted poetry by contemporary Mapuche/Huilliche poets such as Elicura Chihuailaf Nahuelpan (who, in 2020, became the first Indigenous winner of Chile's National Prize for Literature), Jaime Luis Huenún, and Graciela Huinao (the first Indigenous woman to join the Academia Chilena de la Lengua). I include here a selection of verse

Drimys andina (dwarf winter's bark); confocal image

50 μm

by Huenún in *El consumo de lo que somos: muestra de poesía ecológica hispánica contemporánea*. In so many ways, his poetry is, as Jonathan Bate would say, *the song of the earth*. And sometimes Huenún's poems sing in Mapuzugun. In my translation of a translation into Spanish, this would be:

"Inche, Mawiza ñi Pelom,	"I, Light of the Forests,
Witrunko ñi Rayen,	Flower at the water's Source,
ñien kiñe ül	I have a song
pewmatun ñi kewün mew	in the language of dreams
eymingealu."	for you."

During their ceremonies, Mapuche shamans ascend a *rewe*, which Bacigalupo describes as "a step-notched axis mundi, or tree of life, which connects the human and spirit worlds, [allowing them to travel] in ecstatic flights to other worlds". The author says that this sacred structure permits an altered state of consciousness called *küymi* directly linked to *Drimys*: "Mapuche often refer to the *rewe* itself as *foyé* or *canelo*", even though it is often shaped from the wood of another revered tree *triwe* (*Laurelia sempervirens*). According to Mösbach, author of the classic compilation *Botánica indígena de Chile*, the Mapuche consider the Foye tree "a symbol of benevolence, peace, and justice".

The researchers Mariana Cardoso Oshiro and Ivone Antônia de Souza did a "systematic review on the phytochemical, biological, pharmacological and toxicological activities of the species of the genus *Drimys*", shrubs/trees with a wide geographical distribution that are "widely used in Latin American popular medicine for the treatment of malaria, gastric pain, toothache, anaemia", and many other maladies. This overview of existing scientific literature from 1987-2022 was published in Brazil in 2023. Table 1 of their study is a meticulous listing of a compendium of scientific studies demonstrating the medicinal uses of a variety of species of *Drimys* as antioxidant, anti-inflammatory, antimicrobial, antiparasitic, antiviral, and also as an insecticide, an insect repellent, and for use as a bioherbicide. In Chile, a decoction prepared from the bark of the tree is used to treat skin ailments that affect cattle. In Costa Rica, people chew leaves to alleviate toothaches. In Brazil, different species of *Drimys* were shown to have anti-ulcerative, antifungal, and antiviral properties, in addition to significantly reducing triglycerides and total cholesterol. Still other studies demonstrated the efficacy of *Drimys* against neurodegenerative illnesses such as Huntington's disease, and also as a means to keep cancerous cells from proliferating. Importantly, studies also show that *Drimys* is not toxic in humans when it is consumed in moderation. The scientists conclude that future studies are necessary and "will contribute to maximising the therapeutic benefits", which appear to be many, and "minimising the possible risks associated with the use of species of *Drimys*".

above: Traditional Mapuche healer (machi) ascending a sacred rewe in southern Chile in a 19th century photograph

right: *Drymis andina* (dwarf winter's bark); flowers and fruit; Conguillío Park in Curacautín, Chile

below: *Drymis winteri* (foyé); in bloom

Drimys andina

Erythroxylaceae
Erythroxylum novogranatense

Coca

Without a doubt, Coca is one of the most venerated plants of the Americas as a cultural and religious symbol that ensured social cohesion. It also has a plethora of nutritional and medicinal benefits. As Wade Davis puts it: "The chewing of the sacred leaves is the purest expression of Indigenous life".

One of my all-time favourite books is Wade Davis's *One River: Explorations and Discoveries in the Amazon Rain Forest*. It is full of harrowing adventures, heroic feats in the name of science, and a deep respect for traditional Amerindian botanical knowledge as studied by the legendary Richard Evans Schultes and his protégés, Timothy Plowman and Davis himself.

Writing with exemplary eloquence about the importance of *Erythroxylum novogranatense* to the Indigenous groups living in what is now known as Colombia, Wade Davis says: "this was the coca of the thirteenth century Muisca and Quimbaya goldsmiths, the stimulant of the unknown people who carved the monolithic jaguar statues and massive tombs at San Agustín in southern Colombia 1,500 years before Columbus…".

He continues by highlighting the extraordinary cultural significance of this plant: "in the Andes, to use coca is to be *Runa Kuna*, of the people, and the chewing of the sacred leaves is the purest expression of Indigenous life. Take away access to coca, and you destroy the spirit of the people […] The Inca revered coca above all other plants. For them it was a living manifestation of the divine; its place of cultivation a natural sanctuary approached by all mortals on bended knee".

Erythroxylum novogranatense (coca); among the most highly-revered plants in South America

50 μm

Davis's travelling companion in South America, Timothy Plowman (1944-1989), was an accomplished expert on all things related to the genus Erythroxylum, particularly coca, about which he wrote:

"Coca plays a central role in the daily lives of many different groups of South American Indians, not only as a stimulant and medicine, but also as a unifying cultural and religious symbol".

Plowman's research highlights the spiritual value of this plant: "ritual coca chewing enabled shamans and priests to meditate, to enter trance-like states, or to communicate with the supernatural world, even though coca produces slight mental distortions compared to hallucinogenic plants such as *Datura* and *Banisteriopsis* or even tobacco".

For him, coca bridges the geographical diversity (highland/lowland) of the areas in which it is cultivated and unites different Indigenous peoples in its uses: "in both the Andean and Amazonian cultures, reverence for coca is reflected in its widespread use in divination, both for shamanistic healing practices and for predicting the future".

In *Beneath the Surface of Things*, Davis remembers being with the Barasana on what must have been his first trip to Colombia many years ago, and seeing the process of making *mambe* from toasted and powdered coca leaves mixed with the ash of *Cecropia sciadophylla* (the preferred recipe of his Harvard mentor Richard Evans Schultes). He describes walking through the forest the next day: "fortified by a huge wad of *mambe*, I moved effortlessly over rough terrain and, for the first time, felt truly oblivious to the tropical heat". Davis describes with indignant fervour how coca, "a benign and highly nutritious plant, revered today by millions and long celebrated by the ancient civilisations of South America as the divine leaf of immortality", was demonised by the US government and made the focus for eradication in a longstanding War on Drugs at a cost of $1 trillion. He characterises this as a "grotesque failure", that has "robbed us of the promise of one of the most beneficial plants known to botanical science". Davis is adamant that the time is long overdue for coca to recover its sacred legacy so that its therapeutic benefits can be widely available for all people.

opposite: Harvested coca leaves, Yungas, Bolivia
right: Coca plant with fruit, Colombia
below: Cultivated fields of Coca, Yungas, Bolivia

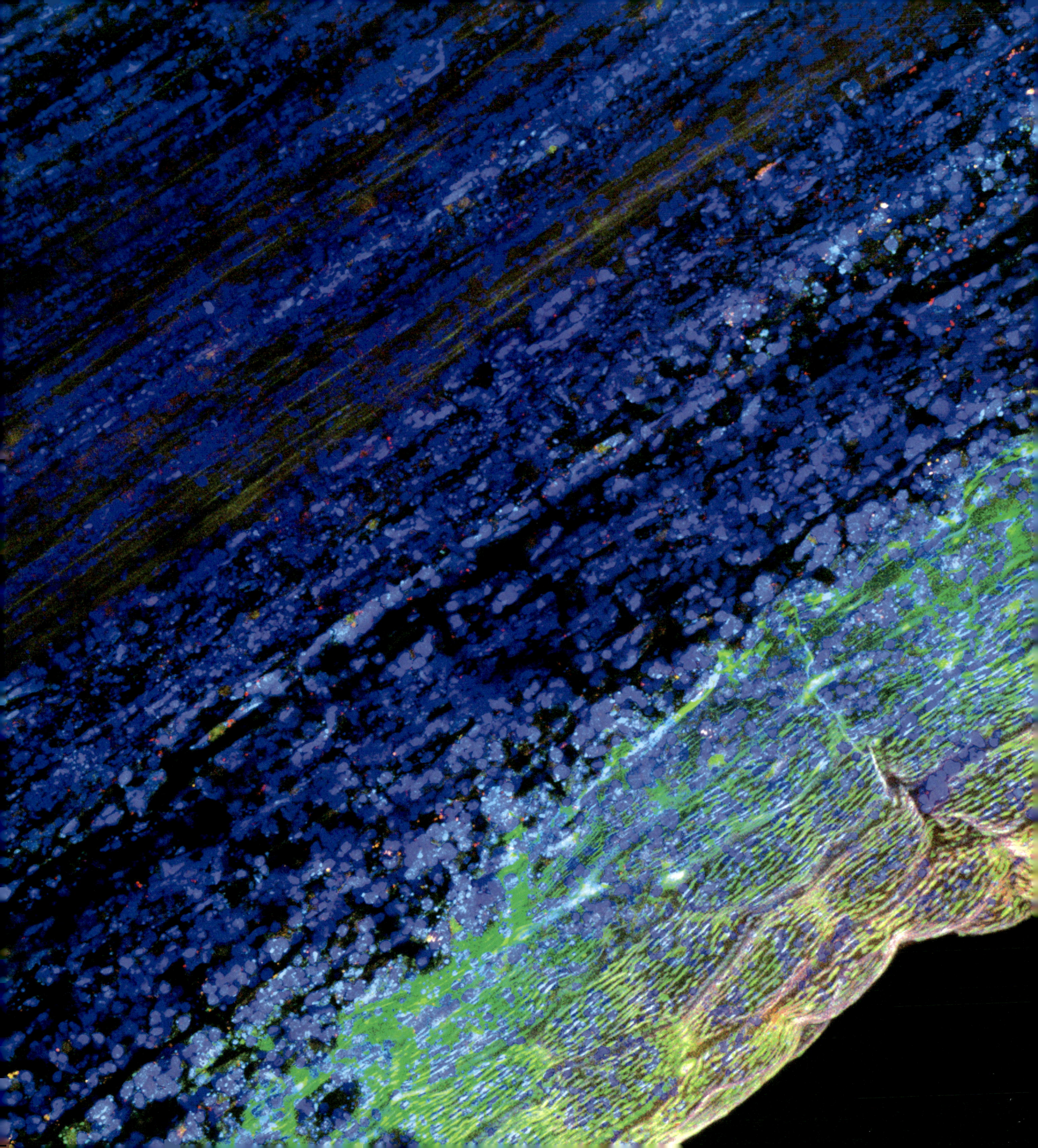

Lythraceae
Heimia salicifolia
Sinicuichi, sun-opener

Known by its common name "sun opener", this plant is sometimes consumed as an infusion, and has the purported entheogenic effect of creating a fleeting golden aura that surrounds all things. It is a perennial herbaceous shrub that grows in the seasonally dry tropical biome with willowlike leaves, small yellow flowers, and tiny seeds contained in ribbed capsules.

Even though there are dozens of names for *Heimia salicifolia* that cut across national borders from Mexico to Brazil, as well as linguistic boundaries that indicate traditional medicinal plant knowledge among a diverse array of Indigenous groups, there is no known pre-Hispanic *ritual* use of this plant. *Heimia* is widely used as a folk medicine (often together with other plants) for ethnogynecological purposes such as infertility, ovarian inflammations and uterine ailments.

The earliest report of *Heimia*, also known as Sinicuiche, as a psychoactive plant is from Juan B. Calderón's 1886 thesis, "Estudio sobre el arbusto llamado sinicuiche". The study is only 27 pages long, and was completed as a requirement for the author to receive his university degree in medicine as a pharmacist. Calderón hoped to substantiate reports of optical effects such as yellow vision and acoustical phenomena after ingesting infusions made from the leaves of Sinicuiche. The young student's 19th century scientific curiosity based on his direct experience with varying, precise amounts of self-administered plant-preparations bears a resemblance to the work of contemporary psychonauts that is available online in the Erowid Experience Vaults, where one can read dozens of earnest "trip reports" for *Heimia salicifolia* conducted over more than two decades. Although Calderón was unable to feel the psychoactive properties of Sinicuiche himself, he does seem to have been successful, as Christian Rätsch puts it, in producing "a morphogenetic field that still exerts itself and continues to develop today". The Erowid experiences range from severe physical discomfort ("This is poison. Do not take it!") to a sublime,

Heimia salicifolia (sun-opener); confocal image

50 μm

euphoric state, albeit of short duration ("Everything was bathed in a soft, wondrous sunshine [...] A truly entheogenic experience if there ever was one!"). Victor A. Reko, in medical journals of the 1920s and 30s, refers to "a magical drink causing oblivion", though it is unclear what, exactly, forms the basis of this conclusion. Professional academic researchers Marvin H. Malone and Ana Rother in their phytochemical and pharmacologic review of *Heimia* published in 1994 also conducted self-experimentation and could find no evidence of the plant's alleged psychoactive properties. Likewise, the best that Jonathan Ott could do in *Pharmacotheon* based on his own personal experience is put *Heimia* on a list of "Doubtful Entheogens". One might ask: is it only a certain variety of the plant that produces visionary qualities, the freshness of the leaves, the quantity consumed, the method of preparing an infusion that might require fermentation? Intuition takes one back to Calderón in the late 1800s and the likelihood that there was some cause for his initial research on sinicuiche, also known as "sun opener". Even so, there is also the ongoing issue of the inconsistent reproducibility of the entheogenic experience proffered by this enigmatic plant that seems to have become a repository for wishful human thinking.

above: *Heimia salicifolia* (sun-opener); flowers
opposite: *Heimia salicifolia* (sun-opener); confocal image

50 μm

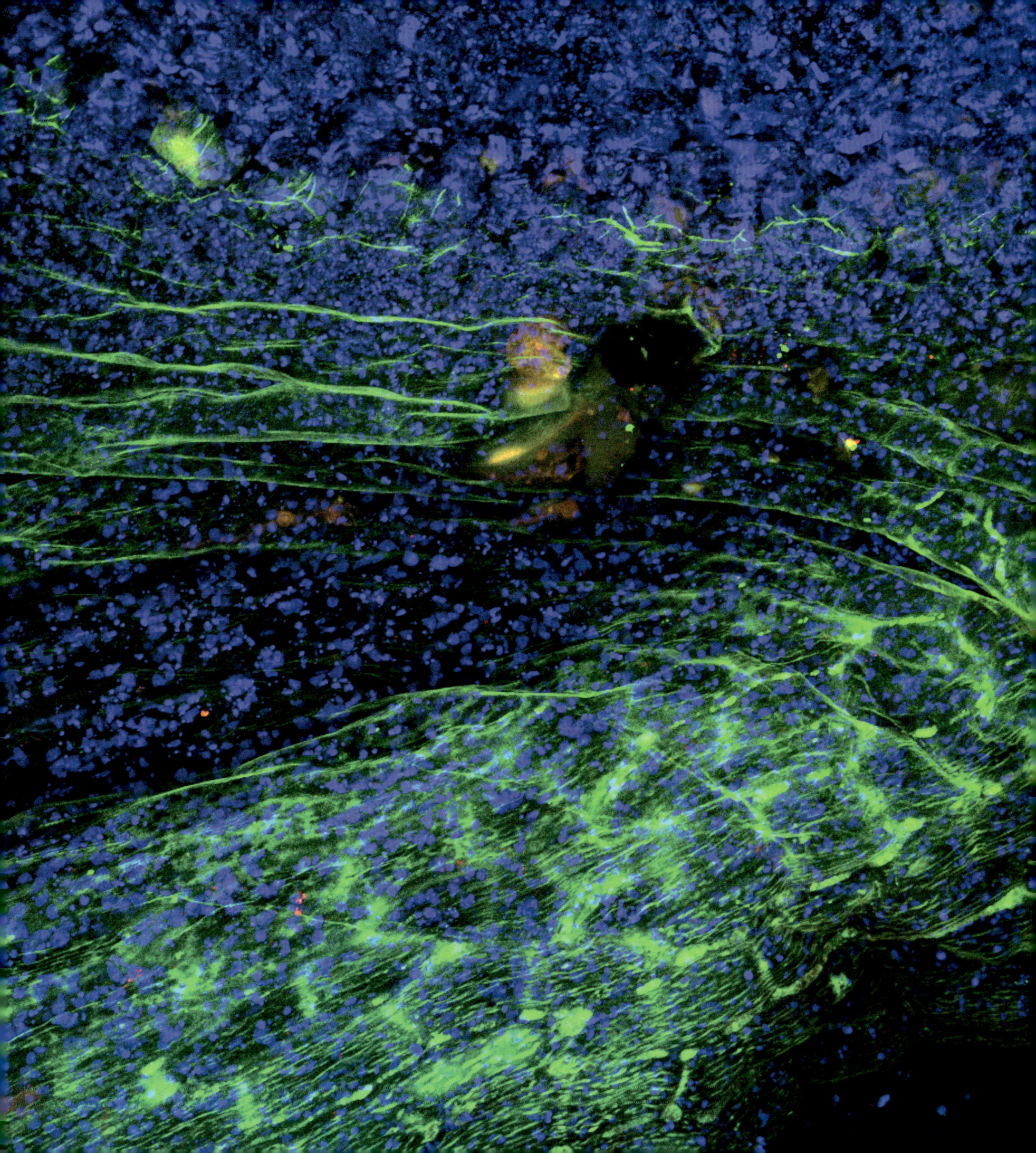

Poaceae

Hierochloe odorata

Holy grass, óhonte wenserákon, sipátsimo, sweetgrass, wicko'bimucko'si, wingaashk

Known as sweetgrass, this perennial grass has multiple names in different Indigenous languages that attest to its widespread presence and significance among Native peoples of North America. Its fragrance is an integral part of purification rituals. The plant is also used for making ornate baskets. Unfortunately, due to loss of wetland habitat and overharvesting, sweetgrass populations are declining.

The name for sweetgrass in Mohawk (Kanien'keha) is óhonte wenserákon and in Cheyenne it is motse'eo. According to Cliff Eaglefeathers and Pete Risingsun, "sweet grass (Motse'eo) is a sacred plant, a gift from Maheo' (God), our Creator of Life. Cheyenne believe life is a spiritual journey with the sacred spirit of Maheo'. We also believe in an invisible spiritual power greater than our own spirit. Sweet grass (Motse'eo) smells like its name, a natural sweet fragrance that invites your spirit into the Circle of Life of Maheo'. We smudge ourselves with the smoke from burning a Motse'eo braid to receive spiritual cleansing and healing from Maheo'. We smudge ourselves and pray for the blessings of Maheo's gifts, that only he can bless us with. These gifts are a clear mind and clean heart from which come patience, keen vision, acute hearing, and thoughtful speech of wisdom".

In the Natural Resources Conservation Service *Plant Guide* prepared by the US Department of Agriculture, there are ample references as to how widespread the use of sweetgrass was and continues to be among Native peoples for the purpose of purification and prayer: "Indian people call sweetgrass the grass that never dies. Even when it is cut, it retains its fragrance and spirit. Today, sweetgrass is used inter-tribally throughout the country. Sweetgrass was used ceremonially by many tribes, including the Omaha, Ponca,

Hierochloe odorata (sweetgrass); confocal image

50 μm

left: Sheila Ransom, Akwesasne artist, working with sweetgrass and splints of black ash (*Fraxinus nigra*) to make ornate baskets at an art fair, Adirondack Museum, Blue Mountain Lake, New York, 2024

above: Robin Wall Kimmerer, author of *Braiding Sweetgrass: Indigenous Wisdom, Scientific Knowledge and the Teachings of Plants*

opposite: *Hierochloe odorata* (sweetgrass); plant

Kiowa, Dakota, Lakota, Blackfeet, Cheyenne, Pawnee, and Winnebago. The Cheyenne, Blackfeet, and Lakota use sweetgrass in the Sun Dance [...] In the northeast, the Ojibwe, Potawatomi, Winnebago, Menominee, Mohawk, Penobscot, Passamaquoddy, and Abenaki made coiled baskets of sweetgrass".

The best work by far on sweetgrass and its relation to Indigenous plant knowledge is *Braiding Sweetgrass: Indigenous Wisdom, Scientific Knowledge, and the Teachings of Plants* by Robin Wall Kimmerer, a Distinguished Teaching Professor at the SUNY College of Environmental Science and Forestry in Syracuse, New York. She is also an enrolled member of the Citizen Potawatomi Nation. Because of her academic background and graduate work in Botany, her perspective combines hard science with Native American traditions, a difficult task to be sure. As she puts it, "getting scientists to consider the validity of Indigenous knowledge is like swimming upstream in cold, cold water". Nevertheless, her book is itself a kind of weaving: "this braid is woven from three strands: Indigenous ways of knowing, scientific knowledge, and the story of an Anishinabekwe scientist trying to bring them together in service to what matters most. It is an intertwining of science, spirit, and story – old stories and new ones that can be medicine for our broken relationship with earth".

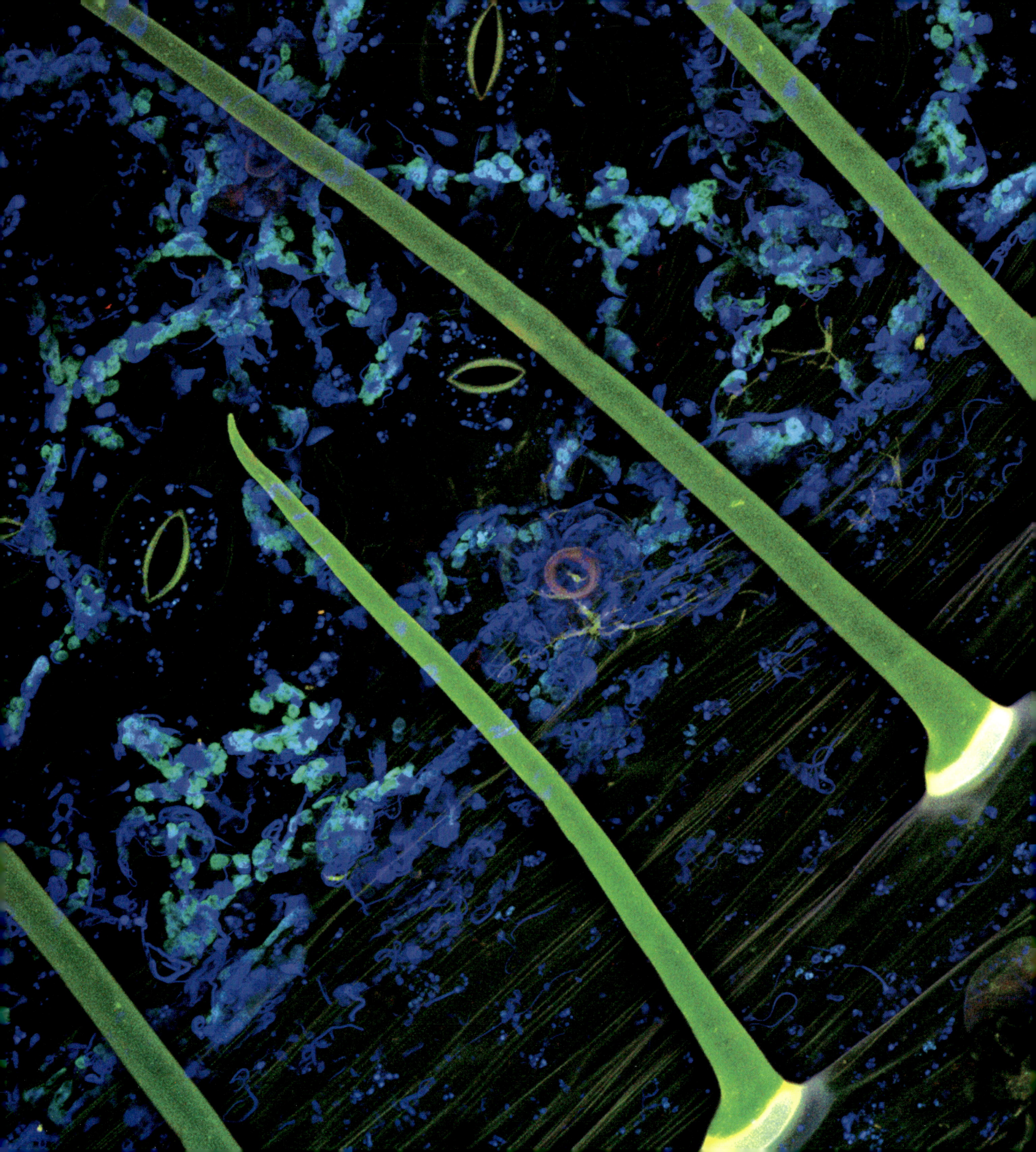

Convolvulaceae

Ipomoea corymbosa

Christmas vine, ololiuhqui, snake plant, x-táabentun

&

Ipomoea tricolor

Badoh negro, heavenly blue morning glory, tlililtzin

Both species of the so-called Mexican morning glories are climbing annual vines with heart-shaped leaves. Flowers of Ipomoea tricolor are heavenly blue, pink, or white, opening in the morning and withering before sunset. Ipomoea corymbosa flowers are white with a yellowish centre. The psychoactive seeds of Ipomoea tricolor, so important for rituals conducted by the Aztec theocracy, are black and elongate-triangular, whilst fruits of Ipomoea corymbosa only have one rounded brown seed.

According to Wade Davis, Albert Hofmann, the inventor of LSD, discovered that "the active principles of ololiuhqui (*Ipomoea corymbosa*) were two indole alkaloids, lysergic acid amide, and lysergic acid hydroxyethylamide, compounds that he already had sitting on the shelves of his lab".

Ipomoea corymbosa (ololiuhqui); confocal image

50 μm

About this and other members of the Convolvulaceae family, Schultes and Hofmann write in their indispensable work *Plants of the Gods*: "as with the sacred mushrooms, the use of the hallucinogenic Morning Glories, so significant in the life of pre-Hispanic Mexico, hid in the hinterlands until the present century".

Fagetti's research on the combined ground seeds of *Ipomoea tricolor* (Semillas de la Virgen) and *Datura stramonium* (San José) used in healing ceremonies in Huajuapan de León, Oaxaca, Mexico is based on fieldwork she carried out there in 2010. The results, which include raw and genuine transcriptions of dialogues between an octogenarian Mixtec preparer of potions and people with illnesses, definitively confirm the persistence of ancestral Indigenous plant-knowledge (even if it is syncretised with certain Christian elements, as was the case in the 1950s with María Sabina and her healing mushrooms).

Fagetti makes it clear that the preparer of the ground seed mixture (ingested as a drink and applied topically) is not so much a healer as a *listener* who tries to understand what the plants have ordered. The trance produced by the seeds along with other powerful applied vegetal materials such as the leaves of *Brugmansia* allows the sick person to understand the origins of the illness and engage in self-healing. The seeds are considered to speak and make the patient speak as well, these two voices coming together as a first person plural ("We") with divine visionary powers.

Although the Amerindian ritual use of *Ipomoea corymbosa* remains shrouded in mystery and the subject of much speculation, García Quintanilla and Eastmond Spencer make significant contributions to understanding the properties of this plant among contemporary Maya midwives in Pixoy, Yucatán who use this plant (that they call *X-táabentun*) containing ergonovine with its oxytocic characteristics to induce childbirth. Their ancestral knowledge allows them to administer exactly the right dose at exactly the right time.

In this same exemplary article, the authors link the mythic narrative associated with *Ipomoea corymbosa* to death and rebirth, a fitting origin for this plant used to bring new life into the world. As Mayan oral tradition has it, there were once two sisters: Uts Colel, who was considered good, and Xkeban, who was viewed as a sinner due to the way she freely lived her sexual life, though her close and loving relationship with all plants and animals was widely known. Xkeban died, and when she was found days later, people discovered that her body exuded a marvellous perfume and that the animals defended her even from the flies. Those who walked with Xkeban's body to bury her also took on her pervasive fragrance. Soon, springing from her grave were the flowers of the first *X-táabentun* plant, *Ipomoea corymbosa*. Xkeban had escaped from the Lords of Death in the underworld and was reborn as an emblem of fertility in the form of the plant that helps women as they give birth. The supposedly good sister Uts Colel is said to have died a virgin and was famous for the pestilent smell that surrounded her always in life.

"As with the sacred mushrooms, the use of the hallucinogenic Morning Glories, so significant in the life of pre-Hispanic Mexico, hid in the hinterlands until the present century."

~ Richard Evans Schultes and Albert Hofmann,
Plants of the Gods: Their Sacred, Healing, and Hallucinogenic Powers

The commercially-available honey liqueur from Casa D'Aristi, Xtabentún, is advertised as "inspired" in an original Mayan drink, but it is no longer made from honey produced by stingless Melipona bees nourishing themselves exclusively on the flowers of *Ipomoea corymbosa*. Is it possible that this honey had psychoactive properties and had ceremonial uses as the basis of an ancient beverage? Were the seeds of *Ipomoea corymbosa* added to the fermented drink of the Lacandon Maya *baalche*'? For now, these questions remain unanswered.

Jan Elferink, a Dutch medical biochemist and researcher of ancient Amerindian ethnobotany, discusses how the Aztecs

Ipomoea tricolor (heavenly blue morning glory); confocal image

50 μm

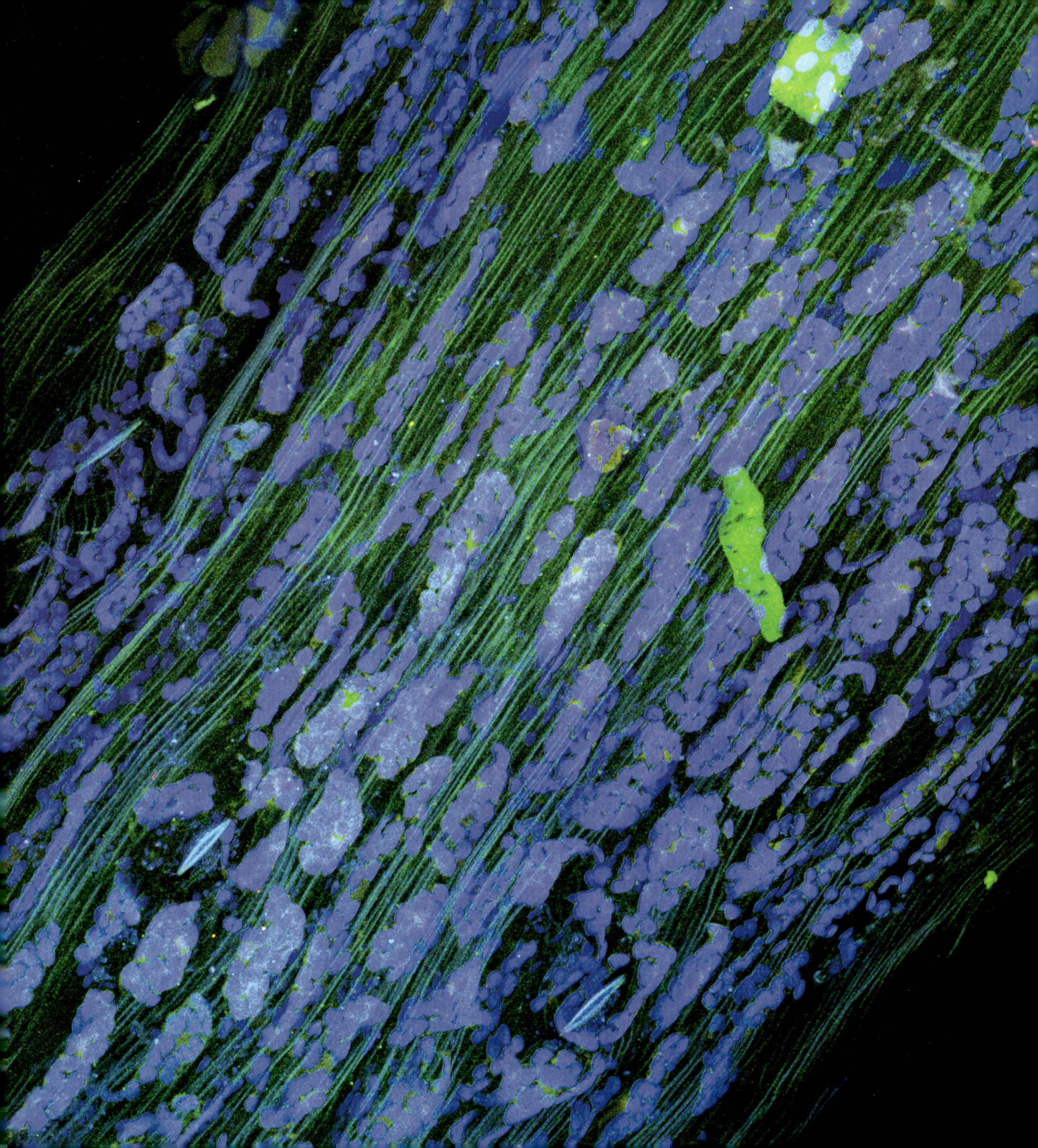

prepared a potent psychoactive bitumen called teotlaqualli, whose main ingredients included ololiuqui, tobacco, and the ashes of different types of charred poisonous animals. The Nahuatl name of this thick black unguent means "divine food", and was used to cover the skin of the priests or even the emperor himself to facilitate the strengthening of the spirit and communication with the gods before performing human sacrifices according to the prevailing religious rites.

Fagetti's work on *Ipomoea tricolor* and *Datura stramonium* as well as the research on *Ipomoea corymbosa* by Alejandra García Quintanilla and Amarella Eastmond Spencer appear in an impressive dossier published by *Cuicuilco: Revista de ciencias antropológicas* on the ritual use of entheogens among a variety of Indigenous groups in Mexico. This issue (53) is a must-read for Spanish-speakers.

In *Mitla: Town of the Souls and Other Zapoteco-Speaking Pueblos of Oaxaca, Mexico*, US anthropologist Elsie Clews Parsons (1875-1941) offers an ethnographic

> *"Contemporary Maya midwives in Pixoy, Yucatán use Ipomoea corymbosa (which they call X-táabentun) containing ergonovine with its oxytocic characteristics to induce childbirth. Their ancestral knowledge allows them to administer exactly the right dose at exactly the right time."*

portrait of a Zapotecan town that is truly remarkable for the author's empathic and meticulously recounted experiences of the economic, political, and religious lives of the families in a predominantly Indigenous community, where the author spent time from 1929 to 1933. In her introduction, Parsons says that "Mitla was undoubtedly an important centre among the ancient Zapotecan peoples" since "its sense of order and organisation, its character of self-possession, its ceremonial elaboration, its *style*, are no short-term developments". The nearly 600-page tome contains a plethora of references to plants used in Mitla for medicinal and ritual purposes, including a plant that Parsons calls *bador*, doubtless a reference to *Ipomoea tricolor*, which Zapotec and Mazatec healers refer to as *badoh negro*. Parsons writes of an invitation she received from a woman in Mitla that, in an enigmatically beautiful way, relates the divinatory morning glory to marking the passage of time, natural cycles of this plant's growth, and its links to people: "'Come to my house!' says Ana on her way home from the mill, her gourd-covered bowl of meal on her head. 'The *bador* which was dry when you were here before is growing now'". Parsons says that Ana's husband is the town's keeper of the potent medicinal plant: "In the yard of Marino Santiago grows a clematis-like vine which is called "spirit children", *bador*; its little boy and little girl appear in the trance produced by eating it and help the sleeper to find what he has lost. They may also tell a sick person whether or not he is to recover". Parsons learns that this is the only plant of its kind growing in Mitla and that the caretaker "sells its leaves or seeds to two or three of the curanderos to administer to patients", meaning, according to Parsons, that the plant represents "a small capital for the family". How is the plant used? Parsons writes that the two curanderos, Agustina and Urbano, "put a leaf on the forehead of one who has lost something and give him thirteen seeds to take in water". The author learned that "after drinking the infusion, the patient, who must be alone with the curer if not in a solitary place where he cannot hear even a cock's crow, falls into a sleep during which the two little ones, male and female, the plant children (*bador*), come and talk". Parsons also relates the following story about divination in relation to *Ipomoea tricolor*: "Don Félix Quero had a herder called José Maria. He lost two cows, and Félix charged him with selling them. That grieved José Maria, so he went to the curandera who gave him the *bador* drink and told him not to be afraid, no matter what came to him, that midnight. The little plant boy came and took him by the hand, saying, 'One of the cows is already meat, the other is about to be killed. Come with me!' He led him in his trance to Tlacolula, to the house of the butcher. The house was closed, but the little plant boy imitated the voice of a compadre, and the butcher let them in. 'There are your animals, hanging on the wall,' said the little plant boy. The next morning the curandera sucked José Maria, for it was dangerous to keep the medicine in him". In her extensive work as anthropologist and ethnographer among Amerindian groups throughout the Americas, Parsons documented the widespread practice of traditional healers sucking illnesses from their patients. She also links the Zapotecan spirit children associated with *Ipomoea tricolor* to the brother and sister in the origin narrative of *Datura* among the Zuñi in the US Southwest.

opposite top left: *Ipomoea corymbosa*, flowers, Tuxtla Gutiérrez, Chiapas, Mexico
opposite bottom left: *Ipomoea corymbosa*, flowers, San Jerónimo, León, Nicaragua
opposite right: *Ipomoea tricolor* (heavenly blue morning glory)

Solanaceae
Latua pubiflora
Arbol de los brujos, kalku-mamüll, latué, latuy

A rare and mysterious entheogen used ceremonially by the Mapuche in southern Chile, this perennial shrub prefers a temperate biome. It seems to grow in all directions, and its violet bell-shaped flowers hang from branches that have long, hard thorns. While latuy, as it is known, is used by traditional healers to dispel evil spirits, the plant's hallucinatory effects put it in the "diabolical" category for Christian missionaries.

Olivos Herreros calls Latué, perhaps the rarest of all psychoactive plants, "the classic hallucinogen of Mapuche ethnology". One researcher translated the name of the plant as "Land of the Dead", perhaps in reference to the isolated region on the mountainous coast of southern Chile (from Valdivia to Chiloé), which is its sole habitat, and which is said to be the place where the dead depart westward with the setting sun for the next life. It is possible to observe this thorny plant, which is in the Solanaceae family, in the Parque Oncol near Valdivia, on the road between La Unión and Hueicolla, and in the Cordillera Pelada near Osorno. According to Rätsch, "for Mapuche shamans, latué is the most important incense for dispelling evil spirits, bad moods, worries, and grief". He also says that the Huilliche "still revere the plant as a shamanic tree, for it brings power, knowledge, and realisation; offers magical protection; and can heal". Latué can also "cause severe delirium and visual hallucinations", the after-effects of which can persist for weeks. Scientific research has established that the main alkaloids responsible for the sedative and hallucinatory effects of *Latua pubiflora* are atropine and scopalomine. According to Sánchez-Montoya et al, "*Latua pubiflora* is indeed used by Mapuche medicine men to induce sedation, to reach a trance state or mystical experience, and also as piscicide". For Capuchin priest and missionary Wilhelm de Mösbach, latué is a "sinister little tree", and one of Chile's "most toxic plants" that "breaks down one's resistance to twisted intentions". The plant is used in traditional medicine to alleviate cramps and rheumatism.

Latua pubiflora (latué); flower

Latua pubiflora (latué); confocal image

50 μm

Lamiaceae
Leonotis nepetifolia
Christmas candlestick, klipp dagga, lion's ear

This annual or perennial upright herb grows in the seasonally dry tropical biome, and has lipped flowers that are tubular, orange, and hairy. These flowers emerge in whorls from green spheres interspersed on the stem like large beads or knots in a cord. Although native to Africa and India, it was widely introduced into Latin America and the Caribbean. New scientific studies point to this plant as a source of new drugs to fight the infectious tropical disease leishmaniasis.

*L*eonotis nepetifolia, although originally from Africa, where its common name is Klipp Dagga, has been naturalised, and can be found throughout the Caribbean and the Americas as well as the Indian subcontinent, where there are more than a dozen common names corresponding to the linguistic diversity of this region, including Lion's Ear. As its square stem would suggest, it belongs to the mint family (Lamiaceae).

Sometimes considered an exotic invasive species, more often it is a highly-revered plant with many therapeutic applications and sacred connotations, especially among Indigenous groups such as the Cora (Náyari) (Mexico) and the Guaraní (Paraguay).

In Trinidad, it is called Shandilay, and is an important folk medicine used to relieve fevers and coughs, as well as the symptoms of diabetes and asthma.

In Spanish-speaking countries, common names include Flor de Mundo (World Flower) and Mota, a moniker that points to the use of its dried leaves and flowers smoked as a marijuana substitute. We are pleased to be able to include confocal images of the flowers with their remarkable juxtapositions of textures around pollen grains, as well as some of the most aesthetically-complex visualisations of trichomes of any plants we have been able to include. The markedly long, thin, straight stem of *Leonotis nepetifolia* has earned the plant names in Spanish associated with walking sticks, canes, and rods that are symbols of religious power: Bastón de San Francisco, Vara de San José,

Leonotis nepetifolia (Christmas candlestick); confocal image

50 μm

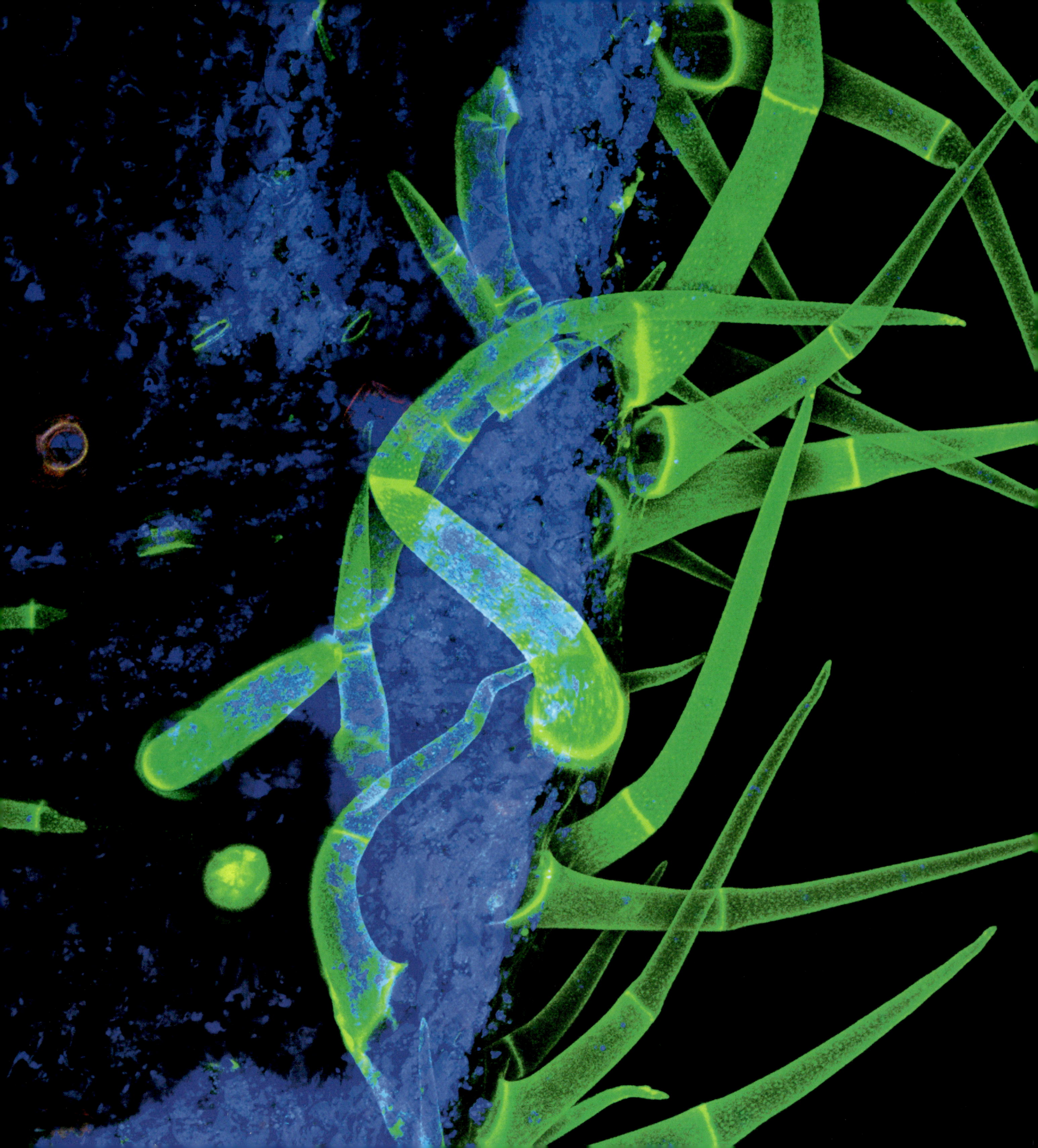

left: *Leonotis nepetifolia* (Christmas candlestick); flower
opposite: *Leonotis nepetifolia* (Christmas candlestick); confocal image

50 μm

and Vara de San Juan. In the common name Bola del Rey (King's Orb), one can see the plant's regal and ornate spherical seed pod. Another common name, Cordón de San Francisco, suggests the way that the stem seems to grow through the centre of the multiple seed pods that, in this sense, resemble a monk's knotted rope cincture. It is from these very prickly spheres that a multitude of beautiful orange furry flowers emerge.

It is interesting to note that in South America (Paraguay, Argentina, Brazil, Bolivia, and Uruguay), non-Indigenous people cultivate *Leonotis nepetifolia* near their homes for medicinal purposes. In the same geographical area, however, the Guaraní people grow this plant, which they call Corazón de San Francisco (Heart of St. Francis), primarily because its tubular blossoms attract hummingbirds, a species that this Indigenous group believes are messengers of lightning, one of their most important deities.

Researchers from Minas Gerais led by Diego Pinto de Oliveira published an overview of *Leonotis nepetifolia* (called *cordão-de-frade* in Brazil) in the journal *Natural Product Research* in 2019. The scientists identified seven flavonoids in this plant (three for the first time) that demonstrated "antileishmanial and anticandidal activities". Previous studies, say the authors, show *Leonotis nepetifolia*'s "analgesic, anti-inflammatory, antioxidant, and antimicrobial activities". Taken together, these qualities make the plant an excellent potential source of new drugs to fight leishmaniasis, a major infectious tropical disease (spread by infected sandflies) that is endemic in Asia, Africa, the Americas, and the Mediterranean region. There are up to 2 million new cases of leishmaniasis each year, and 70,000 deaths worldwide annually.

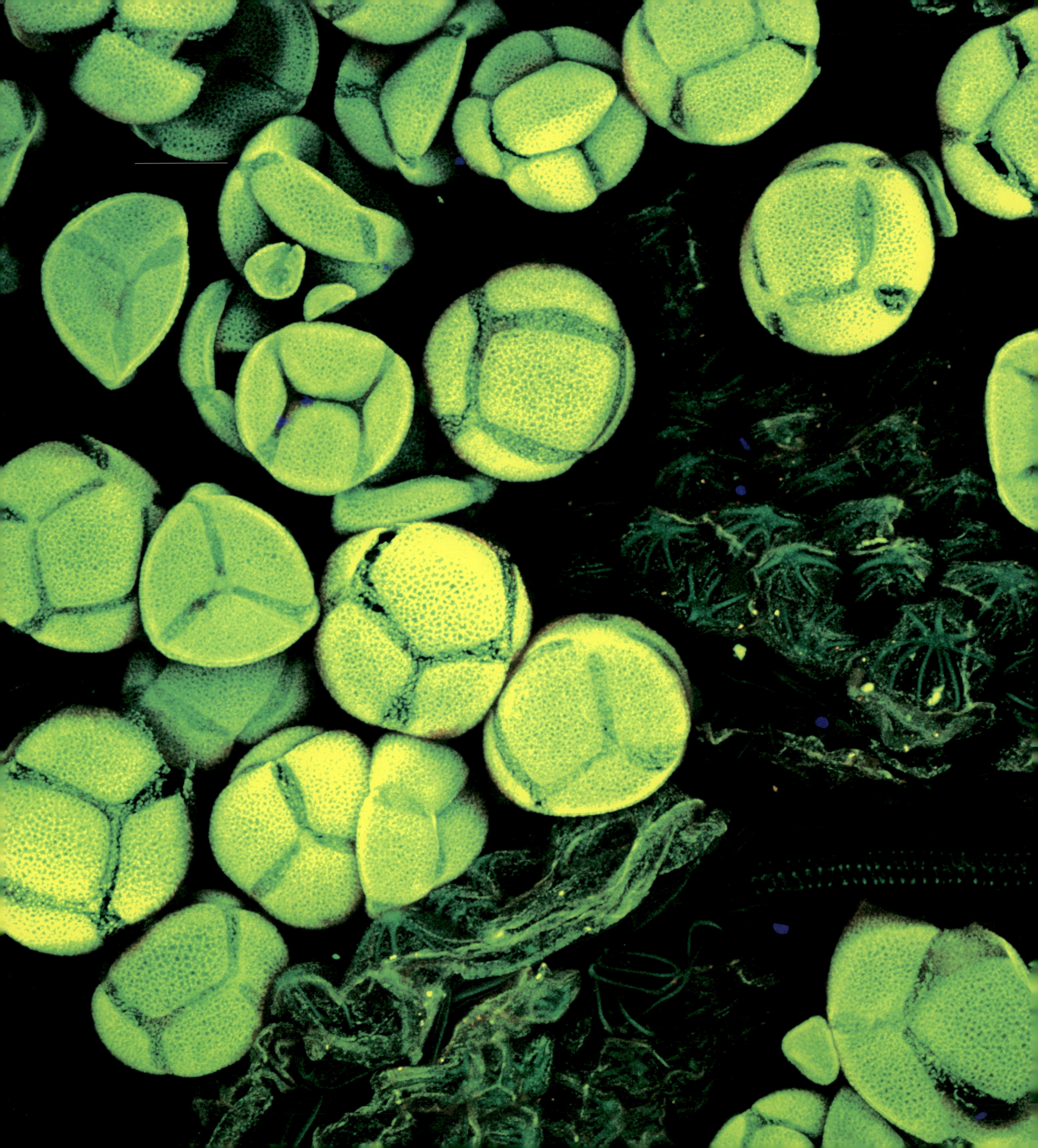

Cactaceae
Lophophora williamsii
Hikuli, peyote

This fleshy, thornless cactus appears as a single head or "button" with little tufts of hair and a lone pinkish flower in the centre of the crown. It is an expert camouflage artist in its desert habitat in southern Texas and northern Mexico. Archaeological evidence confirms that Native peoples, particularly the Wixárika (Huichol), have been eating peyote ceremonially for seven thousand years. It is also a ritual sacrament of the Native American Church, which is comprised of some 250,000 members from federally recognised tribes.

Wade Davis hopes that we always keep in mind a fundamental truth regarding this cactus: "in fact, we now know, based on recent archaeological discoveries, that the native people of Mexico have eaten peyote for seven thousand years".

About that which they characterise as a "divine cactus" used by the Huichol (Wixárika) of Mexico, Stacy B. Schaefer and Peter T. Furst say: "peyote serves as an enculturating force, echoing religious tenets in recurring themes that are transcended to visions, the spoken word through myths and songs, actions in rituals and ceremonies, and beliefs that permeate all levels of individual and collective Huichol consciousness". And in a fascinating article on peyote and women's health, Stacy B. Schaefer concludes that there is a need for further research as to how "the ingestion of peyote alkaloids may influence the production of hormones in the endocrine system via the neurons in the nervous system".

Over the decades, Schaefer has experienced firsthand how the Huichol "have developed and fine-tuned a complex, elaborate worldview that provides members with tools and traditions to heal their bodies, promote fertility, manage healthy pregnancies, and raise their babies".

Lophophora williamsii (peyote); confocal image

50 μm

It is also important to mention in the context of this book the Native American Church (NAC), originally established in North America in 1918, which now has more than 250,000 members who are part of federally recognised tribes, and its use of peyote as a ritual sacrament. The American Indian Religious Freedom Act of 1978 protects the NAC's right to incorporate this cactus as part of its ceremonies.

Attorney Jerry Patchen has written about how "the Native American Church, assisted by ethnologists, ethnobotanists, anthropologists, pharmacologists, and psychiatrists, was the spear point that established the court precedents and legislation that resulted in the legal use of Peyote and ayahuasca for religious purposes in the US".

Of particular interest is the 2023 study by María del Pilar Casado López in which the author identifies a wide range of plants depicted in rock art from Northeastern Mexico, including what appears to be colonies of *Lophophora williamsii*. These discoveries from Nuevo León and Coahuila could be up to 7500 years old. In the words of Casado López, this aesthetic expression of peyote as a ritually-significant plant of power creates a symbolic link "between the ancient first groups of hunter-gatherers and contemporary communities".

above: *Lophophora williamsii* (peyote); northern form
opposite: *Lophophora williamsii* (peyote); confocal image

50 μm

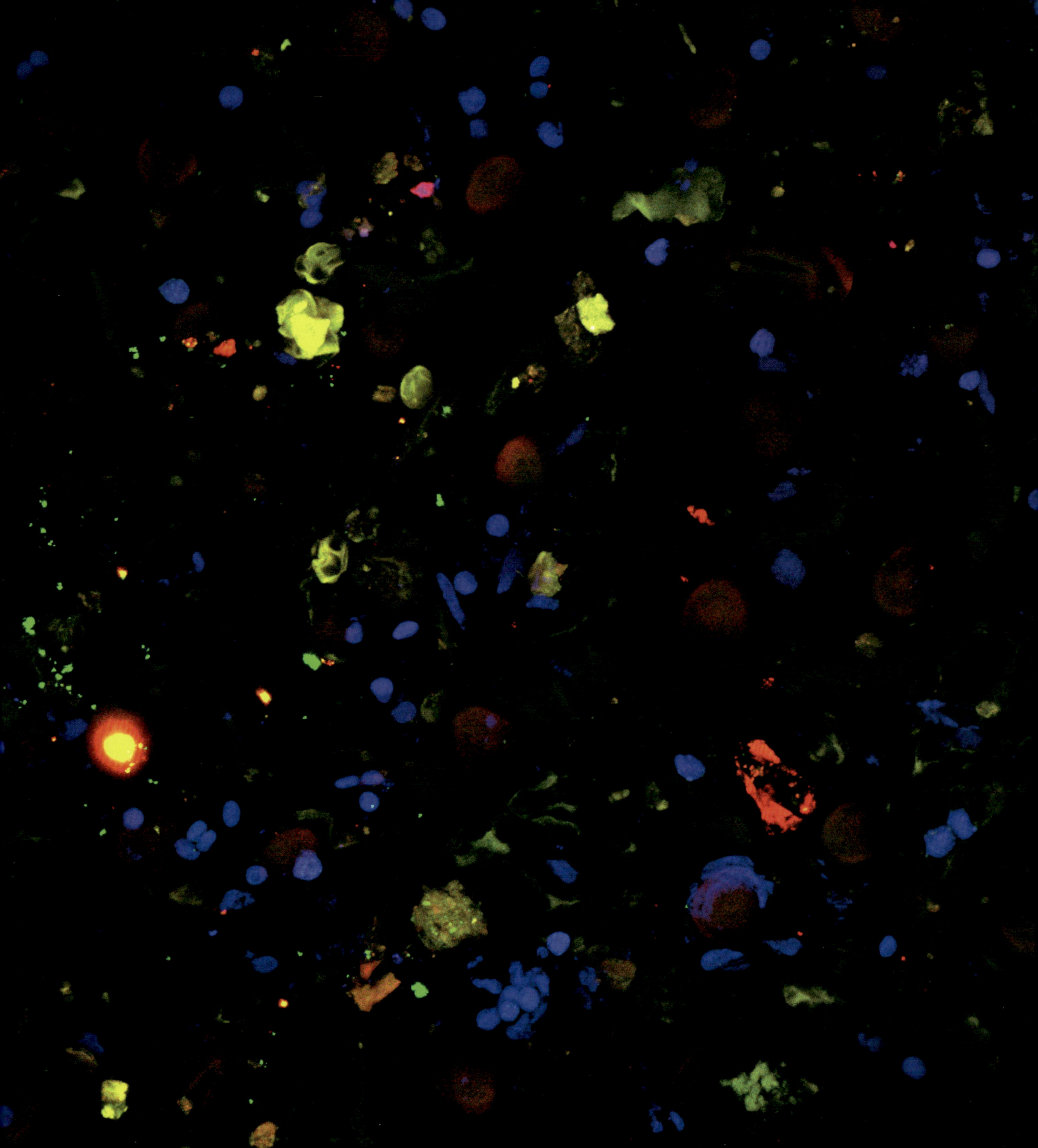

above: Yarn painting on Campeche beeswax and plywood by Wixárika artist José Benítez Sánchez (1938-2009). Here is the explanation handwritten in Spanish on the back of the painting by the artist translated to English:

"The bull is being sacrificed to offer its blood to the sun, the corn, the sacred cactus. The house is the place where the ancestors of the family (parents, grandparents, etc.) reside along with different objects that belonged to them. On the roof, the house keeps the spirits of corn. The shaman is communicating life through his arrows. The serpents are life, the energy of the earth that gives knowledge through the peyote cactus. The shaman transforms himself into the sacred deer, which, at the same time, is fire, solar energy from the sun. The two figures above the corn are offering sacred arrows so they are not forgotten by the gods."

Lophophora williamsii (peyote); confocal image

50 μm

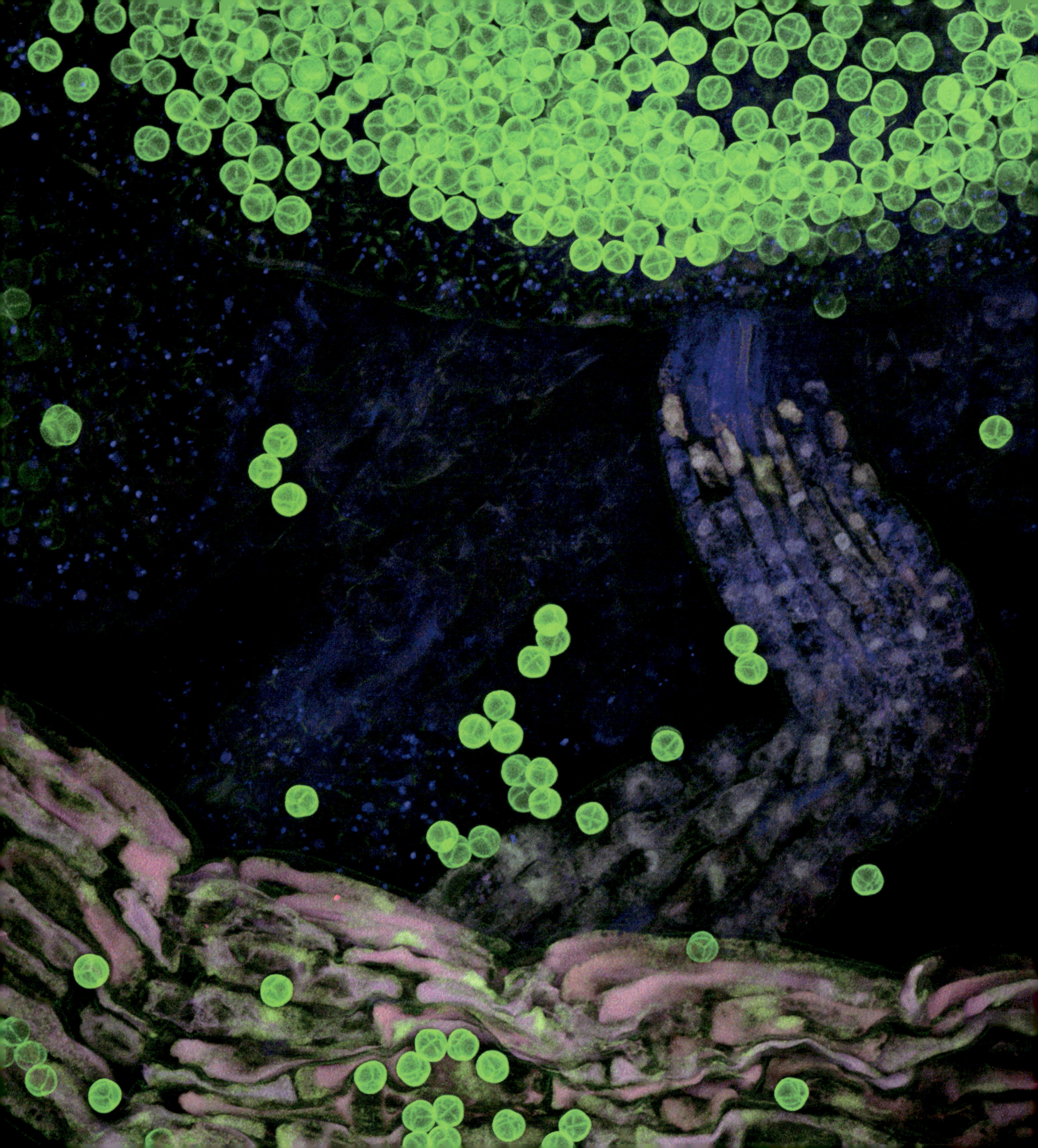

Fabaceae

Mimosa pudica

Dormilona, morí viví, punyo-misa, sensitive plant, touch-me-not

This prickly, perennial, herbaceous plant is known for its alternate bipinnate leaves that are very sensitive to touch. Its pink flower-heads are spherical with many stamen filaments. Researchers such as Monica Gagliano are advancing groundbreaking theories about plant-memory, based on their study of this species that seems to redefine boundaries that traditionally have separated the plant and animal kingdoms, science and the sacred.

Mimosa pudica most definitely has become an important performer in pioneering research on plant-memory. Monica Gagliano considers this species "a strange and extraordinary bridge between two kingdoms of life – the animal and the vegetal, the sensitive and the insensitive".

After a series of rigorous scientific tests conducted in Italy in collaboration with Stefano Mancuso, she concluded that the *Mimosa pudica* plants "had the faculty of memory, and their behaviour was not hard-wired in DNA, but learned".

Gagliano says that *Mimosa pudica* gave her "a sense of elated wonder inspired by something sublime, something magnificent that engenders a deep reverence for life". She considers it a new "sacred" plant.

In *Brilliant Green*, Stefano Mancuso, an authority in the emerging field of plant neurobiology, maintains the following about *Mimosa pudica*: "what's important is the fact that this plant not only has an extremely developed sense of touch, but can distinguish among different stimuli and even change its behaviour, no longer remaining closed once it learns that a stimulus isn't dangerous". Mancuso gives *Mimosa pudica* even more attention in *The Revolutionary Genius of Plants*, considering it "a true botanical star".

Mimosa pudica (sensitive plant); confocal image

50 μm

To make the slides for the confocal microscope, we were fortunate to have access (here at home where I live in upstate New York) to an absolutely perfect flowering specimen of this plant, whose common names include dormilona and touch-me-not.

A group of Nigerian scientists headed by Oluwapelumi E. Adurosakin published an overview of the ethnomedical uses, phytochemistry, pharmacological activities, and toxicological effects of *Mimosa pudica* in 2023 in which they document this plant's hepatoprotective, anti-inflammatory, antimicrobial, wound healing, analgesic, antidiabetic, anxiolytic, antioxidant, anticancer, lipid-lowering, antimalarial, neuroprotective, immunological, diuretic, anthelmintic, anti-ophidian, antifertility, antidepressant, and sedative activities. From this list, it would seem that *Mimosa pudica* is the plant that can do it all! Like other species recorded in this book, *Mimosa pudica* is native to the American continent, but has now increased its geographical distribution, and grows in parts of Africa, Asia, and Australia, where it was introduced.

below: *Mimosa pudica* (sensitive plant); plant and flower
opposite: *Mimosa pudica* (sensitive plant); confocal image
50 μm

Mimosa pudica

Fabaceae
Mimosa tenuiflora
Jurema, jurema preta, tepezcohuite

This scrambling spiny shrub also known as jurema and tepezcohuite flourishes in the dry caatingas of eastern Brazil and arid zones of Mexico. It is the DMT-source in the sacred Afro-Brazilian psychoactive beverage vinho da jurema. Lotions and soaps containing tepezcohuite are marketed globally as effective skincare treatments.

In an overview of the chemical composition and uses of *Mimosa tenuiflora*, Sara Lucía Camargo-Ricalde maintains that, despite an established contemporary tradition of this plant as a Mexican folk medicine for the efficacious treatment of skin problems, burns, and wounds, the author could find no pre-Hispanic references to medicinal uses of tepezcohuite in major sources such as the *Códice Badiano* by Martín de la Cruz, Francisco Hernández´s *Historia de las plantas de la Nueva España*, and the *Códice Florentino* compiled by Fray Bernardino de Sahagún. Furthermore, she was not able to discover any indication that *Mimosa tenuiflora* (previously known as *Mimosa hostilis*) is used currently by different ethnic groups (such as the Zoques, Mixes, Popolocas, Huaves, and Zapotecos) in the geographical region where this tree grows in Mexico. Etymologically, the Náhuatl origin of the tree's name refers to the hardness of its wood, a tree of iron: *tepus-cuahuitl*. Camargo-Ricalde seems to lament the fact that *Mimosa tenuiflora* contains "certain alkaloids" that might be an obstacle to the successful commercialisation of the plant in over-the-counter products. Pedro Cadena-Iñiguez also warns of the potential danger of certain unspecified metabolites in the plant that might produce what he calls "undesirable effects".

Indeed, *Mimosa tenuiflora* (called jurema in Brazil) has been found to be rich in N, N-dimethyltryptamine (DMT), and is utilised in the preparation of a potent drink called *vinho da jurema,* used ritually by contemporary Afro-Brazilian religious groups such as Catimbó, Umbanda, and Candomblé – especially in Pernambuco in northeastern Brazil.

Mimosa tenuiflora (tepezcohuite); confocal image

50 μm

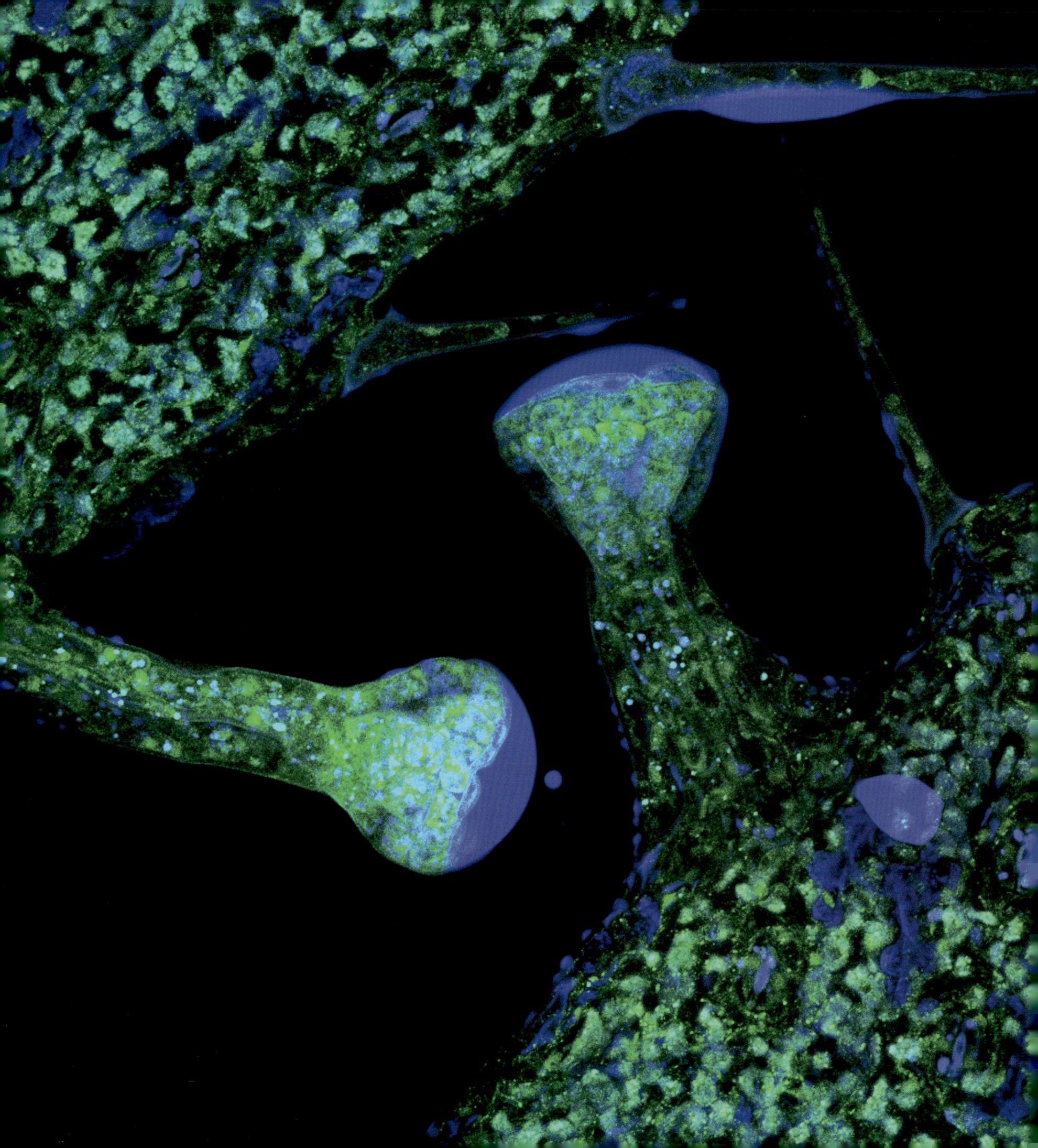

Still to be resolved is the identity of the MAO inhibitor in the ingredients of the Jurema preparation. Is it *Cyperus*, known in Brazil as *dandá* or *junça,* and in Peru as *Piri Piri*? Perhaps the answer can be found in the recent phytochemical studies undertaken by José Jailson Lima Bezerra, among others that are included in the *Microcosms* Bibliography. Christian Rätsch, in his *Encyclopedia of Psychoactive Plants*, documents Indigenous use of *Mimosa tenuiflora* in the eastern Amazon region among the Pancarú, Karirí, Tusha, and Fulnio. Rätsch also mentions that certain Afro-Brazilian ayahuasca cults "venerate Indian spirits (*caboclos*)", among them Cabocla Jurema, "a personification of *Mimosa tenuiflora*".

above: *Mimosa tenuiflora* (tepezcohuite); flowering plant from the Brazilian caatinga
opposite: *Mimosa tenuiflora* (tepezcohuite); confocal image

50 µm

Fabaceae
Neltuma spp.
formerly *Prosopis limensis*

Algarrobo, espino negro, huarango, kiawe, mesquite

Known commonly in English as mesquite, this evergreen and deciduous tree grows primarily in the seasonally dry tropical biome. It thrives in the arid soils that will characterise much of the planet in the future. Often maligned as an invasive species and fire hazard, Neltuma has not been fully appreciated as a valuable source of drought-resistant nitrogen fixers in the legume family, whose seeds are a super food rich in protein and minerals capable of sustaining large human populations in a time of crisis. Because this tree formed the basis of a beneficial ecosystem, the ancient Nazca culture revered it, and made a sacred geoglyph of it on the southern coast of Peru.

Using a term that is part of his Rarámuri (Tarahumara) heritage, Enrique Salmón explains the importance of *iwígara* in the introduction to *Iwígara, the Kinship of Plants and People: American Indian Ethnobotanical Traditions and Science*: "in a worldview based on iwígara, humans are no more important to the natural world than any other form of life. This notion influences how I lead my own life and guides many of my decisions. Knowing that I am related to everything around me and share breath with all living things helps me to focus on my responsibility to honour all forms of life. I carefully consider all living and non-living things when making choices or weighing actions I might take. In short, I see myself as one of many stewards of the land and natural world. I share breath with it, so I endeavour to minister to it with appropriate ritual, thought, and ceremony". Clearly, this is a more complete and profound definition of what might be understood (and perhaps misunderstood) by the more common word "sacred" as it is freely used and even unthinkingly abused in a wide range of cultural contexts. In trying to decide which plants to include in his anthology

Neltuma limensis (mesquite); confocal image

100 μm

left: One of the last fragments of riparian forest dominated by Huarangos (*Neltuma limensis*), and now decimated by logging. This used to be the dominant ecosystem along many of the rivers descending from the Andes and crossing the coastal desert in southern Peru. This tiny remnant is found in Copara (Ica state), not far from the Nazca Lines

below: 1000-year-old Algarrobo (*Neltuma pallida*) in the Bosque de Pomac (Lambayeque state), perhaps the oldest remaining member of its species in Peru of vegetal lives, Salmón says: "before writing this book, I conferred with native plant practitioners, my professional ethnobotanical network, and with close friends. I asked these knowledge holders and wisdom keepers to help me compile a list of plants that are the most culturally relevant to North American native peoples".

There are 80 plant entries in *Iwígara*, a compendium based on a collective sense of respect for specific plants as well as ancestral knowledge that is practical in that it contributes to human wellbeing and survival. Included, of course, is the "revered being" peyote (*Lophophora williamsii*), which Salmón presents in the personalised ceremonial context of the Native American Church: "peyote and the NAC have been credited with saving the lives of thousands of American Indians who needed a path to help them gain a proper relationship with themselves, with their community, and with the spirit world". But Salmón also acknowledges mesquite (*Neltuma* spp.) in his rigorous selection for being an important source of food, fuel, and medicine, as well as a keystone species for desert ecosystems that need to be carefully managed by humans so that, as Salmón puts it, "open mesquite groves in turn encourage native flora and fauna to remain in the area [and] natural diversity returns". Overexploiting in the case of both species has led to serious consequences: placement on the Endangered Species List in Texas in the contemporary case of peyote, and the collapse of the entire ancient Nasca civilisation in coastal Peru when the *Neltuma* forests were cleared and the land was left vulnerable to both flooding and desertification.

Microcosms – Sacred Plants of the Americas seeks to express deep thanks and pay tribute to certain plants as well as their stewards who have been faithful to the spiritual pacts they have sustained with the natural world, and the plant stories they have heard and preserved. Some, but not all, of the species mentioned in this book fall in the category of what Schultes and Hofmann call "Plants of the Gods", due to their psychoactive properties. *Neltuma*, however, known popularly as mesquite, algarrobo, and huarango (among many other names) modifies the definition of what often is considered to constitute sacredness in perhaps an unexpected way: its wood is worthy of the gods. Or would it be more appropriate to say that Pachacamac, one of the most important Pre-Hispanic deities, found a way to reveal himself in a supreme vegetal vehicle capable of conquering time, by means of an exquisitely carved portrait, revered by waves of pilgrims for generation upon generation that has lasted beautifully intact until the present day for more than a millennium?

The Pachacamac Idol, a wooden column more than 2.4m (8ft) tall and 13cm (5in) in diameter, is now a major tourist attraction at the Museo de Sitio Pachacamac located south of Lima, Peru. A team of researchers headed by Marcela Sepúlveda recently conducted tests confirming that the wood is in all likelihood *Neltuma pallida* (a synonym of *Neltuma limensis*) carbon-14 dated from 760-876 CE, which situates the artefact at the height of the Wari Empire in coastal Peru. The scientists also discovered that the idol was painted in at least three colours, including a red derived from cinnabar, a mercury mineral brought from a great distance and reserved for adorning only what is most highly-esteemed, no doubt as a means to highlight the god's spiritual as well as economic and political power. Centuries before the Inca Empire reached its apogee, the Pachacamac Idol was the centre of a major pilgrimage site and an oracle consulted even by the Emperor. Over the course of time, the Pachacamac Idol demonstrated a remarkable ability to adapt syncretically to evolving religious symbolic systems. In an article published in *Archaeology Magazine*, Marley Brown

> *"The Legacy Legumes collectively present an opportunity to build ecosystems in harsh conditions and hold them in a sustained state of abundance that performs ecosystem services while providing sustenance. Together, these keystone organisms constitute an ecological and nutritional foundation below which humanity should strive never to go. We can manage our planetary ecosystems towards abundance and beyond so there will be enough for all."*
>
> ~ Neil Logan, Legacy Legumes: Trees of Renewal and Abundance

cites archaeologist William Isbell of Binghamton University, who says, "I think the radiocarbon date clearly shows that whether the idol represents the principal image of Pachacamac or not, it was there for a long, long time, and participated in a tremendous number of changes that must have occurred on the central coast over those centuries, spanning the Wari Empire, through the Ychsma period, then into the Inca Empire, and through the Inca right to the beginning of the Spanish colonial period".

The seemingly protean identity of Pachacamac links him to the sun and also to the earth in a powerful centre for divination at the highest levels of different successive empires. Peruvian novelist and poet Pedro Favaron, author of essential studies on the spirituality of Amerindian cultures, meditates on Pachacamac in *La senda del corazón* after having survived a devastating earthquake himself on August 15, 2007 while travelling in the Samaca Valley, in the foothills of the Ica River. Favaron writes: "*El manuscrito de Huarochirí*, a fundamental text to approach the Indigenous thought of the Andes, affirms that the *waka* Pachakamaq remains seated in deep meditation. A single movement of his head causes tremors; and it is said that if he were to rise, the whole earth could come to an end. Pachakamaq is the owner of the tremors; it is understood, then, that the tremors are caused by a living and conscious being with whom humans can enter into relationship and ask for mercy. For Indigenous thought, the forces of nature are neither blind nor deaf, but respond to the prayers and respect of human beings".

How did the Pachacamac Idol survive the wrath of the conquistadors? As the story goes, Hernando Pizarro visited its sacred site in 1533 with the intention of entering the sanctum and breaking the idol in front of the priestly caste in charge of the oracle. Ultimately, was it a higher priority for the Spaniards to satisfy their lust for gold as they searched every secret confine of the temple than to destroy the Pachacamac Idol itself? Did the furious, impure, sacrilegious foreign invader simply cast the painted wooden god from its pedestal in the dark chamber bereft of gold? Was the unimaginably hard god in organic form impossible to snap into pieces with ease? These questions remain unanswered. Even so, amazingly, the idol currently on display at the Museo de Sitio near Lima was rediscovered in the Painted Temple's North Atrium in 1938 by Albert Giesecke, who excavated the sculpture from the rubble where it was hidden. Now, as visitors crowd around the Pachacamac Idol in a glass case in Peru, a trembling of the earth in Trujillo, Spain shifts the dusty remains of Hernando Pizarro in his tomb.

opposite: Huarango tree geoglyph, Nazca Lines, southern coastal Peru
overleaf left: *Neltuma* sp. aff. *limensis* (mesquite); confocal image
overleaf right: *Neltuma limensis* (mesquite); confocal image

100 μm

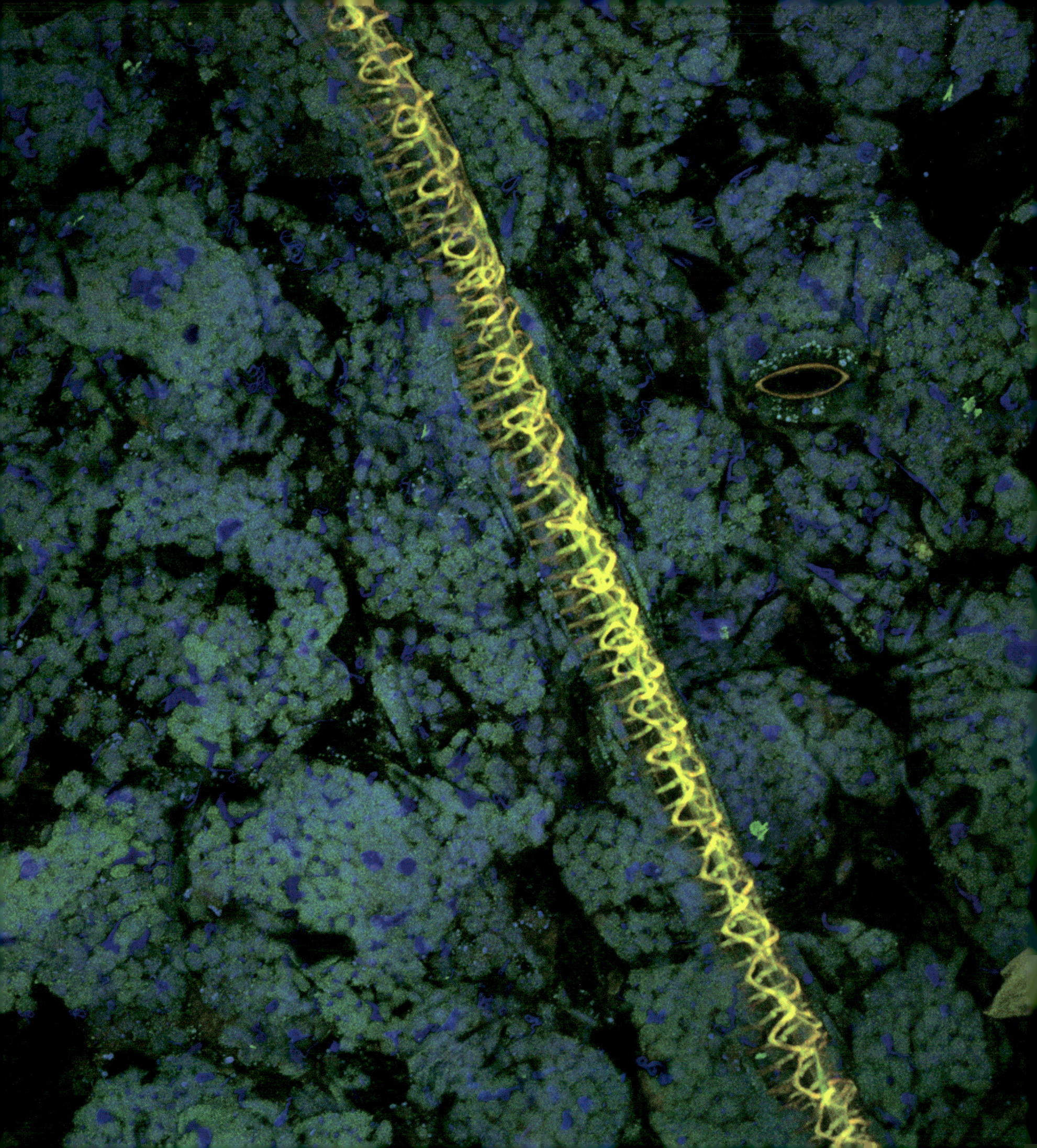

Solanaceae
Nicotiana rustica

Antzil moy, mapacho, piciyetl, sairi, tabaco, tobacco, te, yetl

New scientific evidence demonstrates conclusively that humans were using tobacco at least 12,300 years ago. This annual herbaceous plant from the subtropical biome is native to Peru, but was introduced throughout the Americas. It is a ubiquitous ceremonial plant that is essential for every imaginable ritual, and a foundation of social unification among Indigenous peoples.

Johannes Wilbert's impossibly comprehensive study of tobacco has stood the test of decades: "tobacco in traditional South American societies […] is shown to have played a culture-building role. Functioning as an actualising principle between the telluric and the cosmic, it has served to validate the normative behaviour and to affirm cultural institutions".

Wilbert documents (with a certain amount of chagrin due to his scientific purism) the coexistence of a variety of plants in combination with tobacco: "especially vexing, in this regard, is the overlapping geographical distributions of potential source plants and the simultaneous use of snuffs derived from them within the same region or tribe. Consequently, tobacco snuffing is not always clearly distinguishable from that of other intoxicating materials. Further exasperating the problem is the practice, in some societies, of blending tobacco with *yopo* (prepared from *Anadenanthera*), *parica* (from *Virola*), *coca* (from *Erythroxylum*), or still other substances".

Wilbert confirms the fundamental importance of this plant among a vast range of Amerindian cultures: "in terms of geographic reach and cultural penetration, tobacco has few, if any, rivals among psychotropic plants in pre- and postindustrial societies".

Russell and Rahman agree wholeheartedly: "…regardless of location, the one plant used more than any other was tobacco. Virtually every Amerindian society knew tobacco".

Nicotiana rustica (tobacco); confocal image

50 μm

And so does the major researcher and co-inventor of the term *entheogen*, Jonathan Ott: "tobacco, manifestly, is the fundamental and irrecusable element of American shamanic entheognosia. Virtually no well-known American shamanic inebriant exists independently of some connection with tobacco…".

In an astonishing demonstration of linguistic detective work, Roland B. Dixon documents hundreds of Amerindian words for tobacco used by Indigenous groups from Alaska to Patagonia. His most important conclusion (from 1921) seems to corroborate current research described by Russell and Rahman that the ancestral plants of *Nicotiana rustica* are believed to be *Nicotiana paniculata* and *Nicotiana undulata*, both from North-Central Peru. From his vantage point as a linguist, Dixon affirms the importance of the Quechua word for tobacco still used by Peruvian shamans (according to Francoise Barbira Freedman): "only one case has been found in which a single stem seems to have a wide distribution among unrelated languages, that of *sairi*, for which, however, no extra-American source can be claimed. The situation is, in fact, just what would be expected if tobacco had been known and used by the American Indians for centuries or even thousands of years".

Barbira Freedman reveals amazing details as to how tobacco is essential for nourishing the *yausa*, or *yachay*, the "knowledge-phlegm" that the shaman keeps in his trachea. This phlegm contains darts that hold shamanic power, as well as small animals called *karawa* that include scorpions, spiders, and millipedes received from other shamans as gifts or stolen as they emerge from the mouths of moribund healers. Barbira Freedman says that "without tobacco smoke and also tobacco juice as regular food, these entities become inactive and impotent, not responding to shamans' agentive intentions".

Robert Hall mentions an extremely important idea regarding the ubiquitous nature of this plant-teacher in Amerindian rituals: "the main evidence of antiquity is the pervading holiness of tobacco. It was a sacrifice, a ritual fumigant, a good-will offering, and a sacrament. It was used to seal treaties, friendships, and solemn, binding agreements, to begin war, conclude peace, and legitimise covenants of every description between man and man, between man and the supernatural. Tobacco was used in rites of curing and in rites of human sacrifice".

And because one can never say enough about the enormous significance of tobacco, I was fascinated by the metaphor that appears in this reflection by Glenn H. Shepard, Jr., in an article about his experiences doing fieldwork with Peru's Matsigenka. He was told the following about the tobacco paste called *opatsa seri* that this Indigenous group prepares for shamanic purposes: "when you swallow it, it is like planting a seed in your heart… Each time you take *opatsa seri*, your soul grows like a tree".

Kevin P. Groark is an American Psychological and Medical Anthropologist who teaches at Macquarie University in Sydney, Australia as well as the New Center for Psychoanalysis in Los Angeles, whose research, according to his website, has a "long-term ethnographic focus on the Tzotzil-speaking Chamula Maya of highland Chiapas, Mexico", embedded in what Groark calls "the emergent paradigm of *cultural psychodynamics*". His exemplary article from the *Journal of Ethnobiology*, "The Angel in the Gourd: Ritual, Therapeutic, and Protective Uses of Tobacco (*Nicotiana tabacum*) Among the Tzeltal and Tzotzil Maya of Chiapas, Mexico", is the result of nearly two decades of research and close contact with this particular Amerindian group and their fascinating ethnobotanical practices in relation to tobacco (see also *Breath and Smoke: Tobacco Use among the Maya*). Groark maintains that all forms of tobacco are highly valued by the Maya, though they consider their snuff preparation the most powerful way to benefit from the plant as "a medicine, a stimulant, a protective agent, as well as an intoxicant". He continues by saying that "this mixture, stored and carried in small polished gourds, is the embodiment of an unbroken

> *"Tobacco, manifestly, is the fundamental and irrecusable element of American shamanic entheognosia. Virtually no well-known American shamanic inebriant exists independently of some connection with tobacco."*
>
> ~ Jonathan Ott, Shamanic Snuffs or Entheogenic Errhines

right and below: *Nicotiana rustica* (tobacco); tobacco plant near the ruins of Chavín de Huantar, Peru

tradition of Mayan oral tobacco snuff use spanning more than a thousand years". Groark includes an explanation of the process to prepare the snuff: leaf-collection, de-veining, pounding, addition of admixtures (such as slaked limestone as an alkalising agent), and storage in a tobacco gourd. In considerable detail, the author then describes the intoxicating effects of ingesting this preparation. As one might imagine, there is an entire section on the gourds that are used now as *yavil moy* (tobacco's place/vessel) as well as the ceramic containers used for holding tobacco among the ancient Maya. As a therapeutic powerful substance, tobacco is administered in a variety of ways for the treatment of gastrointestinal ailments, intestinal worms, broken bones, sprains and bruises, tuberculosis, toothaches, gangrene, mange, and boils. Tobacco is also believed to repel evil forces, to blind witches, and also to serve as "one of the primordial foods of the deities, offered to them during fiestas and rituals through proxy ingestion by religious officeholders". One can spit tobacco juice at a coming storm to calm the winds, and toward snakes to paralyse them. Tobacco rubbed on the body can prevent "shock-induced soul loss", and its strong odour facilitates "soul collecting rituals". Groark also mentions that tobacco not only serves in this life, but also after death as a kind of distinguishing merit badge: "frequent use of tobacco snuff is believed to leave an invisible and indelible green stain in the centre of the palm, blessing the user with an afterlife of ease and repose". Chamula syncretic narratives link the Sun-Christ deity's tobacco gourd to Hummingbird, "messenger of the Sun, and protective animal companion of warriors throughout Mesoamerica". Groark ends his study with a discussion of contemporary threats to these traditional uses of tobacco that include "the availability of commercial cigarettes, combined with widespread conversion to evangelical Protestantism". Reading a study of this kind, one feels closer to a preferred "Indigenous research paradigm", even if, as is often the case, the Amerindian perspective is mediated by a Euro-American academic, who lives for extended periods of time in close proximity with the community being studied.

Nicotiana rustica

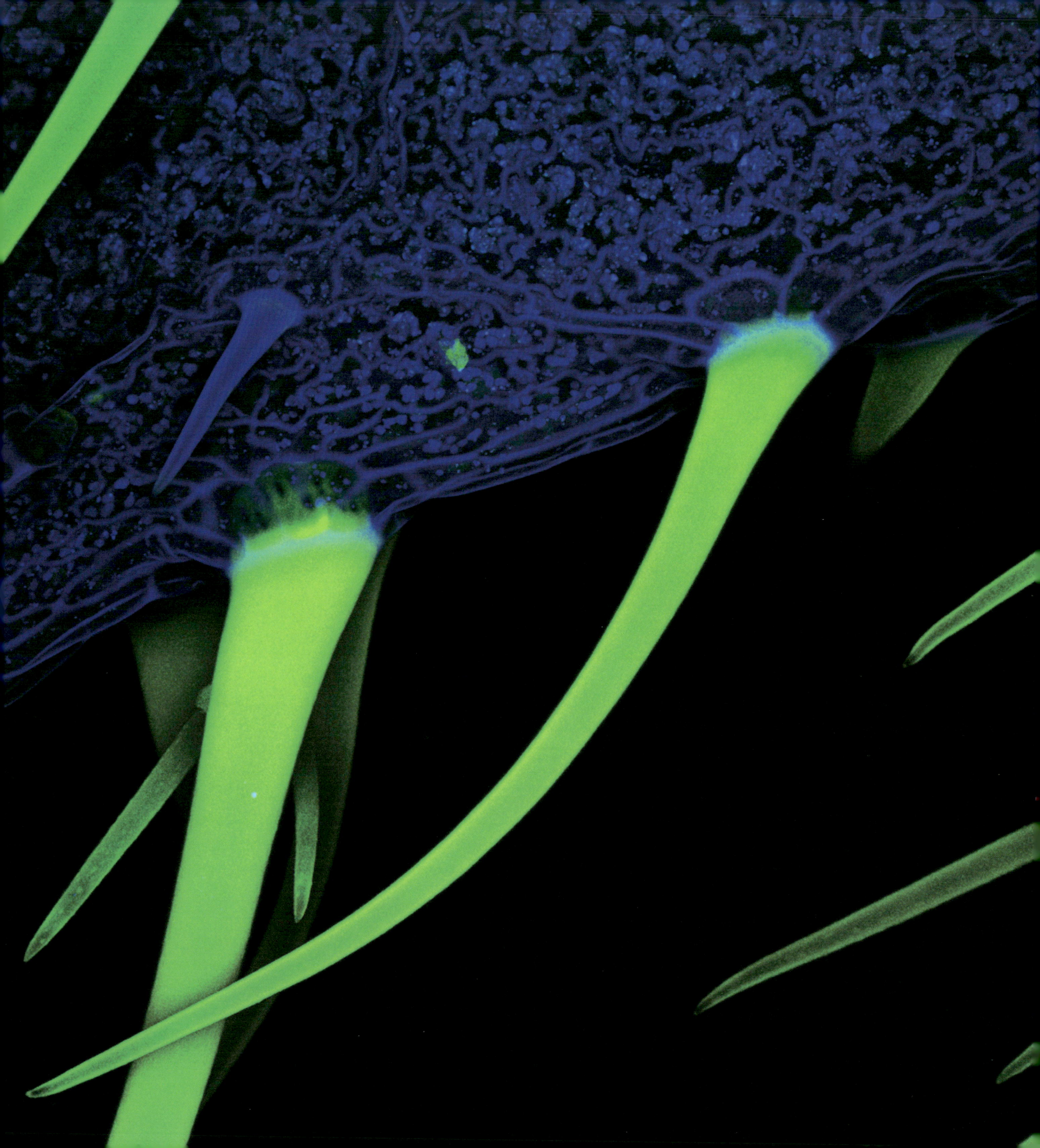

Sapindaceae
Paullinia cupana
Guaraná

This evergreen shrub with scandent branches thrives in the humid lowland tropics of Brazil. Its ground black seeds contain four times more caffeine than any other plant in the world, and are used in a wide variety of energy drinks. New scientific research indicates that it contains phytochemicals that can be used to treat cognitive disorders such as Alzheimer's disease.

This plant, whose common name is guaraná, is held sacred by the Sateré Maué tribe in the Brazilian Amazon and has been cultivated by them for hundreds of years. Containing more caffeine than any plant in the world (2-5 times as much as coffee), guaraná was traditionally their revered source for boosting energy while traversing the rainforest and waging war against distant enemies.

Several myths (collected by Medeiros Marques) in which this plant is a protagonist share a narrative in which a beloved boy (the son of Onhiámuaçabe) is murdered by a jealous, malevolent god. His eyes are buried by the villagers, and his afflicted mother waters them with her tears. As a form of consolation, a benevolent god gives the people Guaraná: a wild plant grows from the left eye and a domesticated plant from the right.

The fruit does indeed resemble a human eyeball when it ripens and the red skin opens to expose an underlying white mesocarp that splits to reveal a black, iris-like seed.

To date, the most comprehensive overview of the pharmacological properties of *Paullinia cupana* is the one published in 2019 by a team of Brazilian researchers headed by Leila Larisa Medeiros Marques. Guaraná is cultivated and processed almost exclusively in Brazil, the production of which is estimated to be at least 4300 tons per

Paullinia cupana (guaraná); confocal image

50 μm

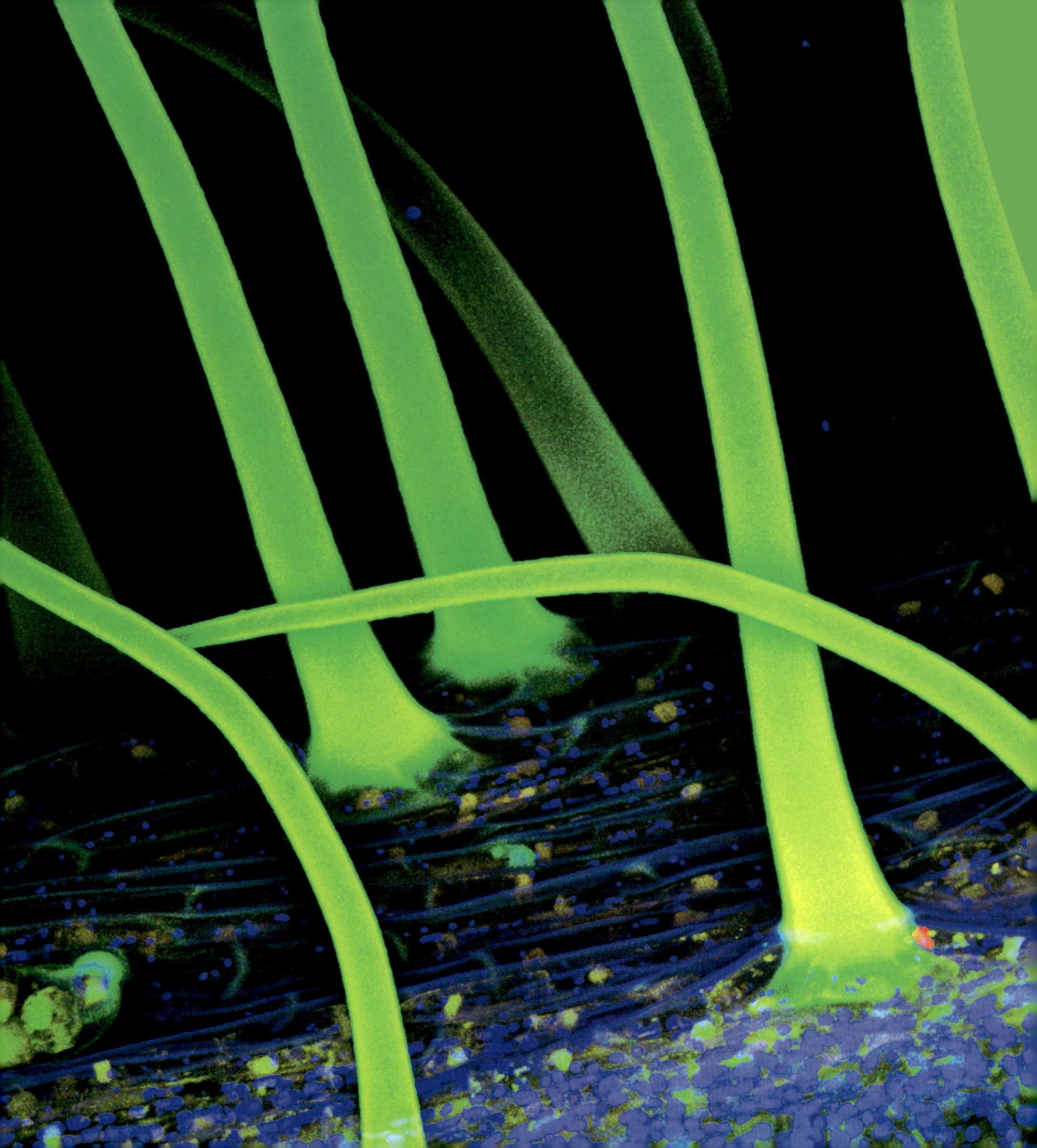

year. The authors affirm that commercial demand for guaraná seeds for use in soft drinks and by pharmaceutical and cosmetic manufacturers has risen steadily in recent years. They also mention the fungal disease anthracnose as a threat to the production of *Paullinia cupana*, and discuss research that is being done regarding disease management. The review also documents research indicating that the high level of caffeine in energy drinks containing guaraná can produce adverse effects such as anxiety and disorders of the central nervous system, and can also exacerbate epileptic seizures. Guaraná is being used by many as a way to control weight gain, and also has been found to have neuroprotective properties that may make it a "promising source of phytochemicals that can be used as an adjuvant therapy in the management of cognitive disorders such as Alzheimer's disease". Research conducted by the authors of the overview showed that guaraná extract has "both anxiolytic and panicolytic effects".

above: *Paullinia cupana* (guaraná); fruiting plant, Brazil
opposite: *Paullinia cupana* (guaraná); confocal image

Confocal image: 50 μm

Rosaceae

Polylepis incarum

Lampaya, queñua

This gnarled and contorted tree grows in the high-elevation regions of the tropical Andes. Its thick trunk has exfoliating, multi-layered, red bark that protects it against low temperatures. Despite extensive reforestation programs, it is highly endangered due to its vulnerability to rural populations seeking firewood and building materials. According to Incan creation narratives, these sacred trees were the ancestors of humanity.

We are pleased to include a contribution here by Ben Kamm, founder of *Sacred Succulents*, an organisation that specialises in rare and endangered plants from the Andean countries, which Ben has explored extensively.

Polylepis, the Andean Progenitors of Humanity, by Ben Kamm

It is estimated that forests of *Polylepis*, also known as Queñal, Kewiña, Keñua, once covered over 20% of the Andes up to over 5182m (17,000ft). These forests were slowly cleared over millennia, massacred by human activity over the last 500 years, and are now reduced to almost nothing. There's an island (Titi'kaka – Aymara name; Isla del Sol in Spanish) in an improbable high mountain lake, more of a small freshwater ocean, stretching between snow-blanketed mountains that touch the heavens. The north side of this island, where the sun shimmers and scintillates on the shy, gentle waves lapping rocky shores, where toadlets hop about, having recently emerged from the lake's primordial waters (in whose depths, among the waterweeds and ancient offering made of stone, ceramic, silver and gold, there lurks the toadlet's close kin, the saggy-baggy aquatic

opposite and overleaf: *Polylepis incarum* (queñua); confocal image

50 µm

frog). Further up the shore, past dwarf Baccharis shrubs and columnar, gold-spined *Trichocereus* cactus, there is a crag of rock, where, like amphibians from the water, the first people emerged blinking, sunstruck and in awe at the beauty of the bright world stretching out before them. The founders of the Incan dynasty were Manco Capac and Mama Ocllo, and this site is where the *Pillkukayna* was built, the so-called Temple of the Sun. The first to greet them, like welcoming kin, were small gnarled trees; rarely taller than the height of three men, their twisted and sprawling trunks flaking sheets of thin bronze-red bark, their dense canopy of white furred, green leaflets, dangling clusters of small yellow-green flowers. These tree-kin gave them welcoming shelter from the elements and nourishment from the fruits, birds, and beasts which populated their forests, as well as fresh water from the crystalline springs which sprang from their roots, medicine from their multifold leaves and ever-shedding bark, utility and fire from their hard, dense wood. Perhaps, these very *Polylepis* trees *are* those first ancestors, Andean primogenitors of humanity. Very few of these trees remain, the nearby island of the moon

> *"Polylepis may well be the most emblematic tree genus of the central and northern Andes. It often occurs in an otherwise treeless landscape, forming the highest forests in the Western Hemisphere at elevations of over 4800m (15,700ft)."*
> ~ *Tatiana Erika Boza Espinoza and Michael Kessler*

is now blanketed with Eucalyptus – a favourite, along with Mexican pine, of NGOs which actually pay locals to reforest the Andes with these foreign species. None of the populations of *Polylepis incarum* on Isla del Sol can be considered forest, only relictual stands. The same is mostly true for *Polylepis incarum* growing on the edges of Titicaca and for much of the other 26 species of this unique high-altitude tree distributed along the Andes, from Venezuela all the way to Córdoba in central western Argentina. Except in the highest elevations, watersheds, and steep or inaccessible mountain slopes, the mark of contemporary man and his beasts is heavy in most areas where the tree still occurs. *Polylepis* are amongst the most enchanting trees I have ever encountered with their contorted trunks and peeling bark, not to mention their rebellious nature. This is a tree that actually dares to grow above the treeline. It has extremely hard wood that is excellent for construction and firewood. It is used medicinally for lung and kidney issues, the bark chewed for oral health. Additionally, it is a source of beige, pale-pink and green dyes. *Polylepis* were considered sacred during Incan times and associated with the ancestors: forests were venerated and protected. Propagation and reforestation of *Polylepis* is essential for sustainable development in the Andes. A eubiotic species, *Polylepis* forests are known to harbour the highest biodiversity of any ecosystem in the high Andes.

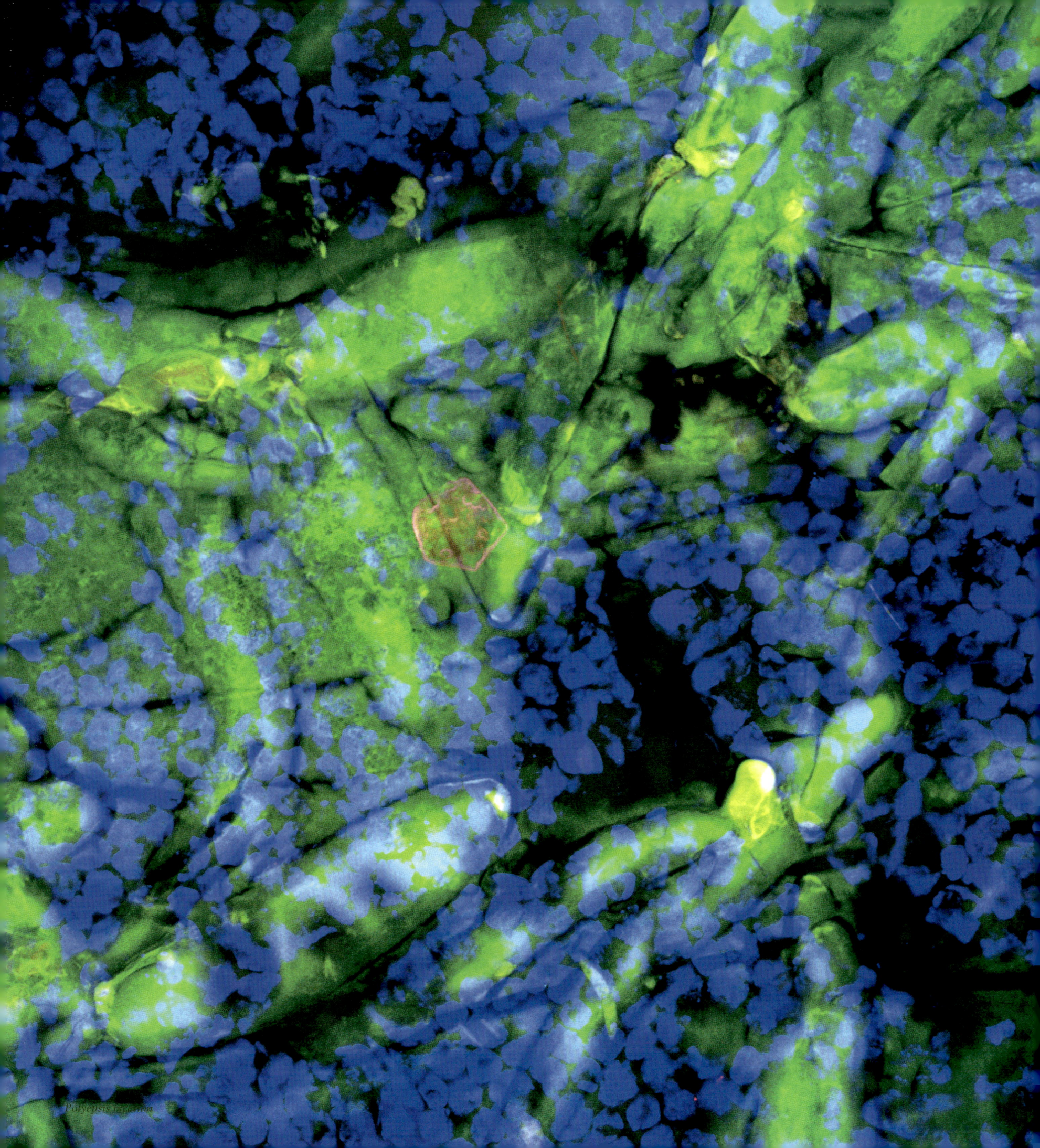

above: *Polylepis australis* (queñua); remnant habitat, Los Gigantes heights, Sierras Grandes, Córdoba, Argentina
left: *Polylepis tarapacana* (queñua); forest, lower dark band, Apu Sajama, Bolivia
opposite top: *Polylepis reticulata* (queñua); trunks and understory, Papallacta, Ecuador
opposite bottom left: *Polylepis lanata* (queñua); forest, Río Lope Mendoza, Cochabamba, Bolivia
opposite bottom right: *Polylepis incarum* (queñua); Isla del Sol, Titicaca, Bolivia

Polylepsis incarum

Hymenogastraceae
Psilocybe cubensis

Di-shi-tjo-le-rra-ja, hongo de San Isidro, magic mushroom, ndi xijtho, niños santos, teonanácatl

The so-called "magic mushrooms" produce the psychoactive compounds psilocybin and psilocin, and were used ritually by traditional healers such as María Sabina in her Mazatec community. New scientific studies are demonstrating the clear potential of this fungus for helping patients struggling with depression and end-of-life anxiety. Psilocybe cubensis is closely related to Psilocybe mexicana, called teonanácatl by the ancient Aztecs.

All multicellular forms of life, including plants, animals, and fungi, evolved from eukaryotic cells. A more accurate title would identify this key psychoactive fungus as a thallophyte, and the flowering vascular plants as angiosperms. At any rate, plants and fungi live in symbiosis, by means of mycorrhizal associations that facilitate the secretion and transportation of chemicals.

As Ralph Metzner has written about the highly-repressive practices during the colonisation of the Americas by Spain, "the suppression of the visionary mushroom cult by the Spanish clergy was effective and complete". This is certainly in keeping with the ongoing inquisitorial spirit perpetuated by hypocritical contemporary anti-drug laws that severely limit research on fungi and plants that have many undeniable benefits for healing, especially in the field of psychiatry, at a time when, worldwide, as a result of the pandemic, we are facing the most serious mental health crisis since World War II. But the ritual use of mushrooms for healing persisted secretly for centuries in remote parts of Mexico, as Alvaro Estrada writes in *María Sabina: Her Life and Chants*.

In June 1955, the US mycologist R. Gordon Wasson was granted permission by Mazatec healer María Sabina who lived in Huautla de Jiménez, Mexico to attend and to document one of her ceremonies in which she chanted and healed the sick after ingesting

Psilocybe cubensis (magic mushroom); confocal image

50 μm

the divine mushrooms. He published articles with dramatic photographs about his profound experiences in *Life* and *Life en Español*.

Three years later, he recorded one of María Sabina's *veladas* (nocturnal vigils) in its entirety. The publicity resulted in a destructive onslaught of foreign "seekers of God". María Sabina later told an interviewer: "from the moment the foreigners arrived, the saint children lost their purity. They lost their force; the foreigners spoiled them. From now on, they won't be any good. There's no remedy for it". In a retrospective essay from 1976, Wasson laments being "held responsible for the end of a religious practice in Mesoamerica that goes back far, for millennia". "I fear", he continues, "she spoke the truth, exemplifying her wisdom. A practice carried on in secret for centuries has now been aerated and aeration spells the end".

In *How to Change Your Mind: What the New Science of Psychedelics Teaches Us About Consciousness, Dying, Addiction, Depression, and Transcendence*, Michael Pollan describes taking some potent *Psilocybe azurescens* mushrooms that he found in the Pacific Northwest with the guidance of Paul Stamets, a leading expert on psilocybin species: "dusk now approaching, the air traffic in the garden had built to a riotous crescendo: the pollinators making their last rounds of the day, the plants still signifying to them with their flowers: *me, me, me*! In one way I knew this scene well – the garden coming briefly back to life after the heat of a summer day has relented – but never had I felt so integral to it. I was no longer the alienated human observer, gazing at the garden from a distance, whether literal or figural, but rather felt part and parcel of all that was transpiring here".

Stamets himself places these same ideas into a global environmental context: "Psilocybin mushrooms carry with them a message from nature about the health of the planet. At a time of planetary crisis brought on by human abuse, the Earth calls out through these mushrooms – sacraments that lead directly to a deeper ecological consciousness and motivate people to take action".

Paul Stamets is collaborating with Giuliana Furci, the Founder and CEO of the Fungi Foundation on an important project called Historias y Memorias Mazatecas, which seeks to preserve the cultural heritage of the Mazatec people. Thus far, over the last two years, work has focused successfully on conserving and restoring historical artifacts and textiles as well as videos and photographs of leading Mazatec healers, including María Sabina. A secure, climate-controlled space has been constructed to protect the contents of the archive which was built over a lifetime by Renato García Dorantes. The collection is now curated by his son, Inti García Flores, a Mazatec historian and secondary school teacher in San Mateo Yoloxochitlán. Future plans, for which fundraising efforts are well underway, include the construction of a museum (and cultural centre) so that these materials and this new space can benefit the Mazatec community.

The Fungi Foundation has been instrumental in promoting what it calls the FFF Initiative, which "elevates fungi's conservation status by advocating for their inclusion in international laws and policies, promoting the term *Funga* alongside *Flora* and *Fauna*". It also seeks to ally its work with the global Rights of Nature movement. To this end, the Fungi Foundation emphasises Indigenous cosmovisions and ancestral relationships to Nature and understands Indigenous people as stewards of the genetics as well as the knowledge associated with medicinal plants and fungi. In an interview with Dennis McKenna on the Brainforest Café podcast series, Furci discusses how giving legal beinghood to mushrooms can accelerate habitat protection in that mushrooms are specific to their host symbionts. She reminds listeners that, unlike plants and animals, fungi cannot be removed from a particular habitat. A team of researchers headed by Sara de la Salle from the Department of Psychiatry at Montréal's McGill University and Hannes Kettner from the Centre for Psychedelic Research at Imperial College London published an article in *Scientific Reports* in 2024 analyzing the results of the pioneering work being done in Canada with regard to the use of psilocybin

> *"Our collective duty as psilocybin allies is to create networks of collaboration that benefit the Commons while respecting Indigenous traditions, whose histories go back hundreds, if not thousands, of years."*
> ~ Paul Stamets, Psilocybin Mushrooms in Their Natural Habitats

to treat anxiodepressive symptoms in patients with life-threatening illnesses. Legal pathways to obtaining access to "magic mushrooms" on the grounds of compassion began in 2020 and has reached perhaps 100 Canadian patients. The researchers "conducted a prospective longitudinal survey which focused on Canadians who were granted Section 56 exemptions for legal psilocybin-assisted psychotherapy". Data gathered from the small number of participants accepted for this formal evaluation suggest "significant improvements in anxiety and depression symptoms, pain, fear of COVID-19, quality of life and spiritual well-being" among most patients.

above: Mazatec healer María Sabina
right: *Psilocybe cubensis* (magic mushroom)

Psilocybe cubensis

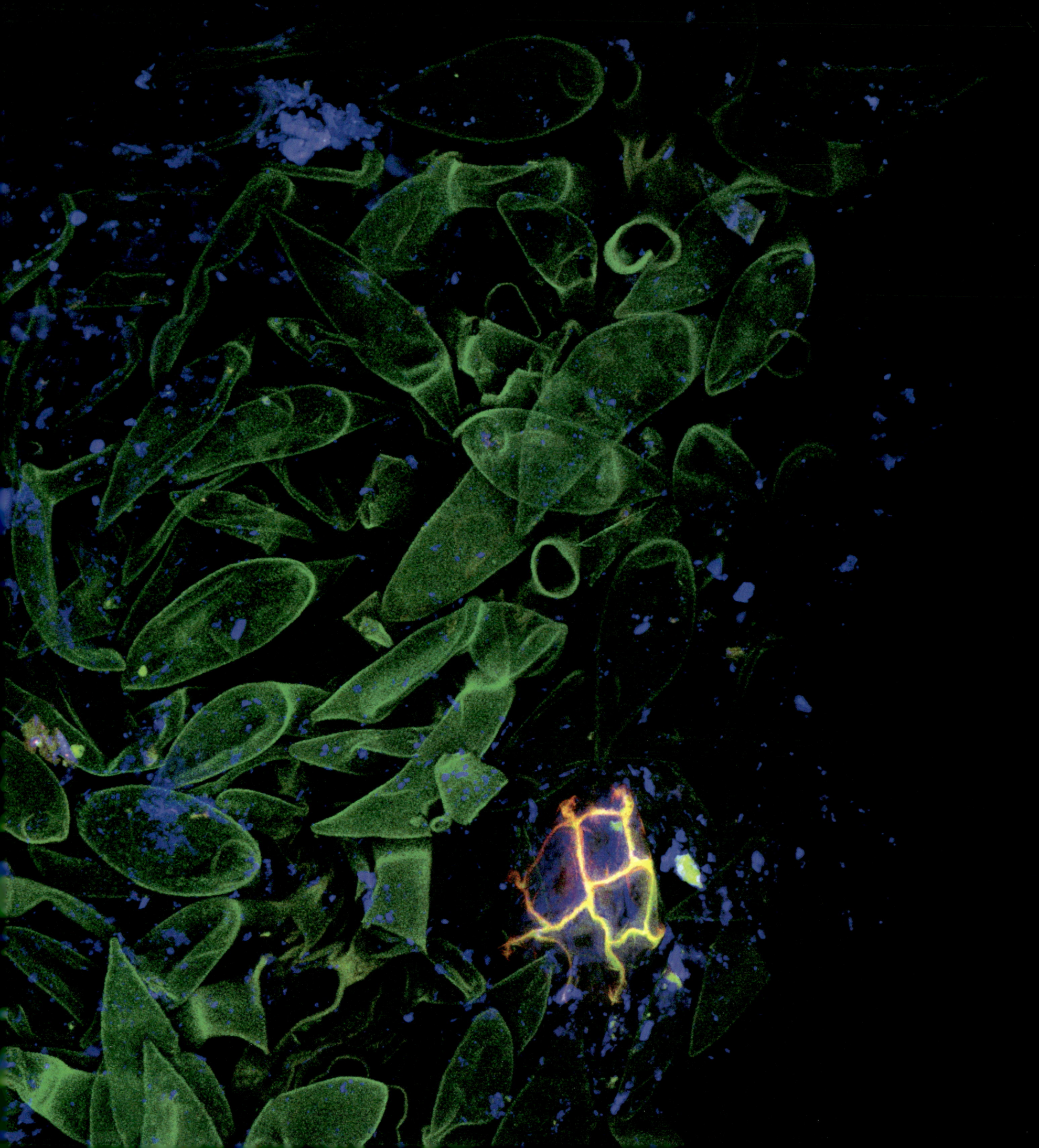

Lamiaceae

Salvia apiana

Bee sage, buffalo sage, California sage, pellytaay, white sage

Most commonly known as white sage, this annual shrub grows in southern California and northwest Mexico. Its whitish evergreen leaves are covered with hairs that trigger oil glands, producing a strong fragrance when rubbed. Due to overharvesting of this revered plant that plays an important role in purification rituals, Native peoples have requested that non-Natives stop using it.

White sage is an important and sacred plant for Native Americans. It provides a source of food and medicine for the Kumeyaay, who burn its leaves in a sweat-house to purify out toxins associated with illness. In living spaces, its leaves are burned as a form of fumigation. The seeds of the white sage can be toasted, ground up, and used as a main ingredient for a meal called *pinole*; the young stalk is peeled and eaten; and the leaves are used to remedy cold and flu.

A major review of *Salvia apiana* in the journal *Planta Medica* in 2022 sought to summarise the biological activities of this plant as a phytomedicine with great potential that has exhibited antioxidative, antimicrobial, and cytotoxic properties. The authors mention the ethnomedicinal and religious importance of *Salvia apiana*, called *khapsikh* by the Chumash Indians of the coastal chaparral region of southern California: "white sage is deeply rooted in tribal culture as an apotropaic herb [and] is believed to have a great power of cleansing the spirit, restoring its balance, drawing a blessing upon people, or even carrying the prayers to God". *Salvia apiana* was also used by traditional Chumash healers together with *Datura meteloides* in social initiation ceremonies for children. Additionally, white sage is an important bee-food, and the unusual morphology of its flowers indicate its co-evolution with large insects, especially *Xyclopa* bees.

Salvia apiana; plant

Salvia apiana (white sage); confocal image

50 μm

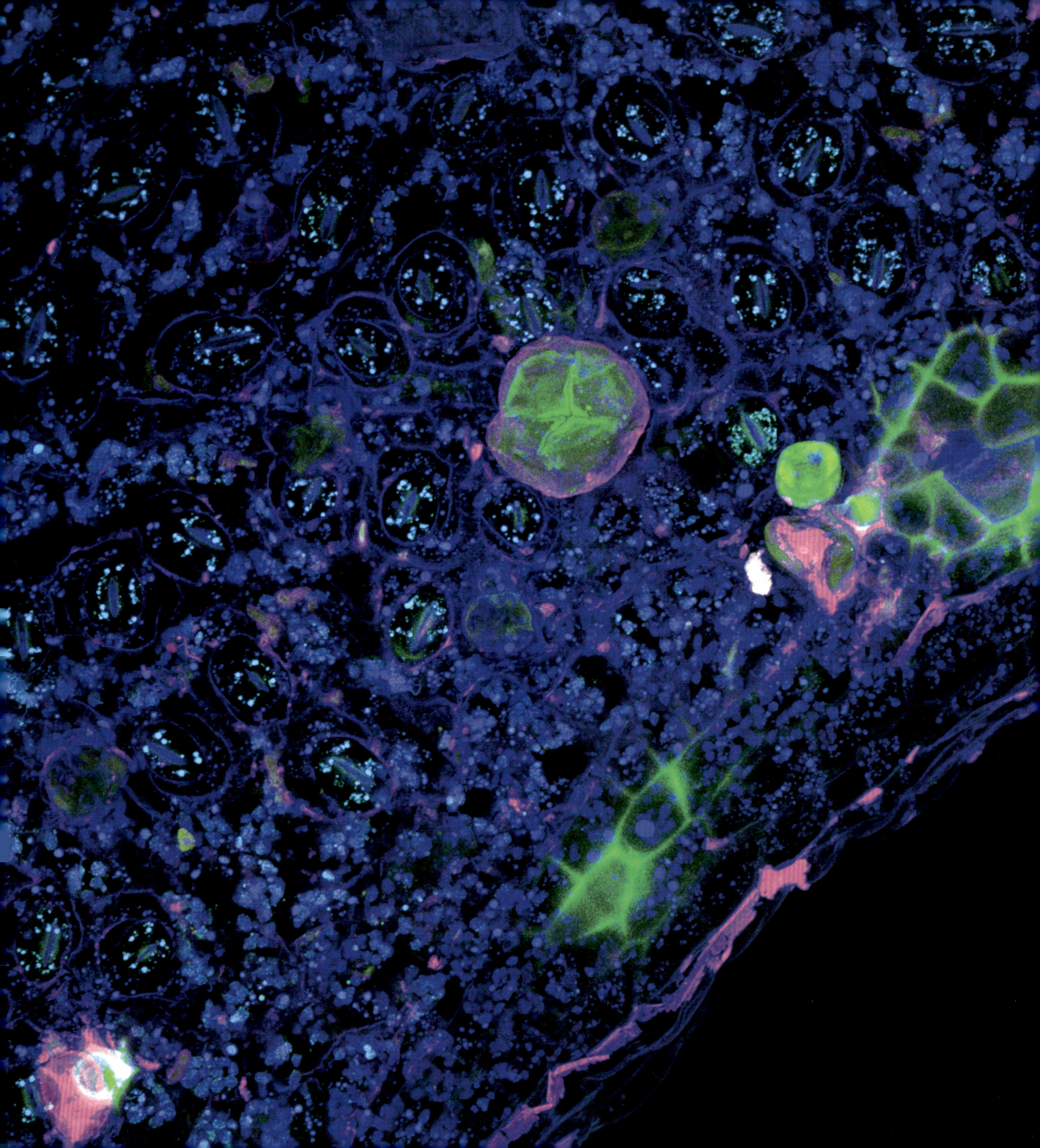

Lamiaceae

Salvia divinorum

Diviner's sage, hierba de la pastora, ska pastora

This evergreen herbaceous perennial native to Oaxaca, Mexico is a member of the Mint family. Known by its common name, ska pastora, fresh leaves are used by traditional Mazatec healers in a ritual context to treat a variety of infirmities ranging from vaginal diseases to cocaine addiction to bronchitis. New research indicates that the terpenoid salvinorin A contained in the plant has potential for the treatment of drug dependence without being addictive itself.

The most comprehensive overview of *Salvia divinorum*, a member of the mint family, was published in the *Journal of Ethnopharmacology* in 2013 by a team of researchers headed by Ivan Casselman.

Their article "concentrates on the investigation of *Salvia divinorum* over the last 50 years including ethnobotany, ethnopharmacology, taxonomy, systematics, genetics, chemistry, and pharmacodynamic and pharmacokinetic research".

In the ethnobotanical section, the authors link traditional uses of the fresh leaves of this plant to Mazatec shamanism in Oaxaca, Mexico, where the plant is used as a palliative for patients near death. Similar approaches are being explored for more effective hospice care in the United States, Canada, and Europe. Men and women Mazatec healers undergo apprenticeship training with three plants: the leaves of *Salvia divinorum*, the seeds of *Ipomoea tricolor*, and *Psilocybe* spp. mushrooms.

"Initially", say the authors, citing work published by Leander J. Valdés, "trainees ingest increasingly large doses of *Salvia divinorum* leaves which show them the way to heaven, where the initiated learn from the tree of knowledge".

With regard to the chemistry of the plant, Casselman's team of researchers confirms that "it is the diterpene salvinorin A that is responsible for the bioactivity in *Salvia divinorum* and which are also considered to be potential lead compounds in pharmaceutical research".

Salvia divinorum (diviner's sage); confocal image

50 μm

In their introduction to a study of *Salvia divinorum* published in *Journal of Pain Research*, Mexican researchers Ulises Coffen and Francisco Pellicer cite traditional uses of this plant to treat "inflammatory conditions and pain, such as headaches, gastrointestinal (GI) problems, or rheumatism", in addition to, among the Mazatecs, "insect bites, eczema, candidiasis, cystitis, and menstrual cramps, and even depression or alcohol addiction". In their conclusion, the authors affirm that "the experimental evidence supports the fact that *Salvia divinorum*, salvinorin A (SA), and their analogues decrease the pain induced by neuropathy and inflammation".

Portuguese scientists directed by Andreia Machado Brito-da-Costa published an extensive article on the pharmacokinetics and pharmacodynamics of *Salvia divinorum* in the journal *Pharmaceuticals* in 2021, in which they study the psychological, physiological, and toxic effects of this plant, and its bioactive compound, the neoclerodane diterpene salvinorin A. As the authors point out, "unlike the other naturally occurring hallucinogens, salvinorin A is a terpenoid that does not have nitrogen atoms in its molecular formulae". The article also provides detailed statistical analysis of recreational use of *Salvia divinorum* in the United States, Canada, and Europe, as well as the plant's global legal status. The authors give considerable attention to forensic techniques for the detection of salvinorin A in products containing *Salvia divinorum*. The researchers maintain that "the short-term effects of *Salvia divinorum* vary widely from person to person, and include modification of visual perception, hallucinations, out of body experiences, altered states of self and reality, dizziness, light headedness, disorientation, mood and somatic sensations, confusion of senses (e.g., hearing colours or smelling sounds), dysphoria, and increased vigilance". The scientists also highlight that "the therapeutic potential of salvinorin A for the treatment of drug dependence [such as cocaine addiction] comes from the drug's capacity to decrease dopaminergic activation and extracellular DA levels". In their conclusion, the authors state that "it is noteworthy that the drug [salvinorin A] seems to induce tolerance without displaying abuse potential nor dependence". Their future research goals seem to be focused on the possibilities of creating analogues of salvinorin A that would not produce what they call undesirable "psychotropic side effects" in their hopes of developing "opiate analgesics with a better safety profile".

One finds quite a different approach in Ana Elda Maqueda's chapter "The Use of *Salvia divinorum* from a Mazatec Perspective" from *Plant Medicines, Healing and Psychedelic Science*, edited by Beatriz Caiuby Labate

top: *Salvia divinorum* (diviner's sage); plant
above: *Salvia divinorum* (diviner's sage); flowers
opposite: *Salvia divinorum* (diviner's sage); confocal image

50 μm

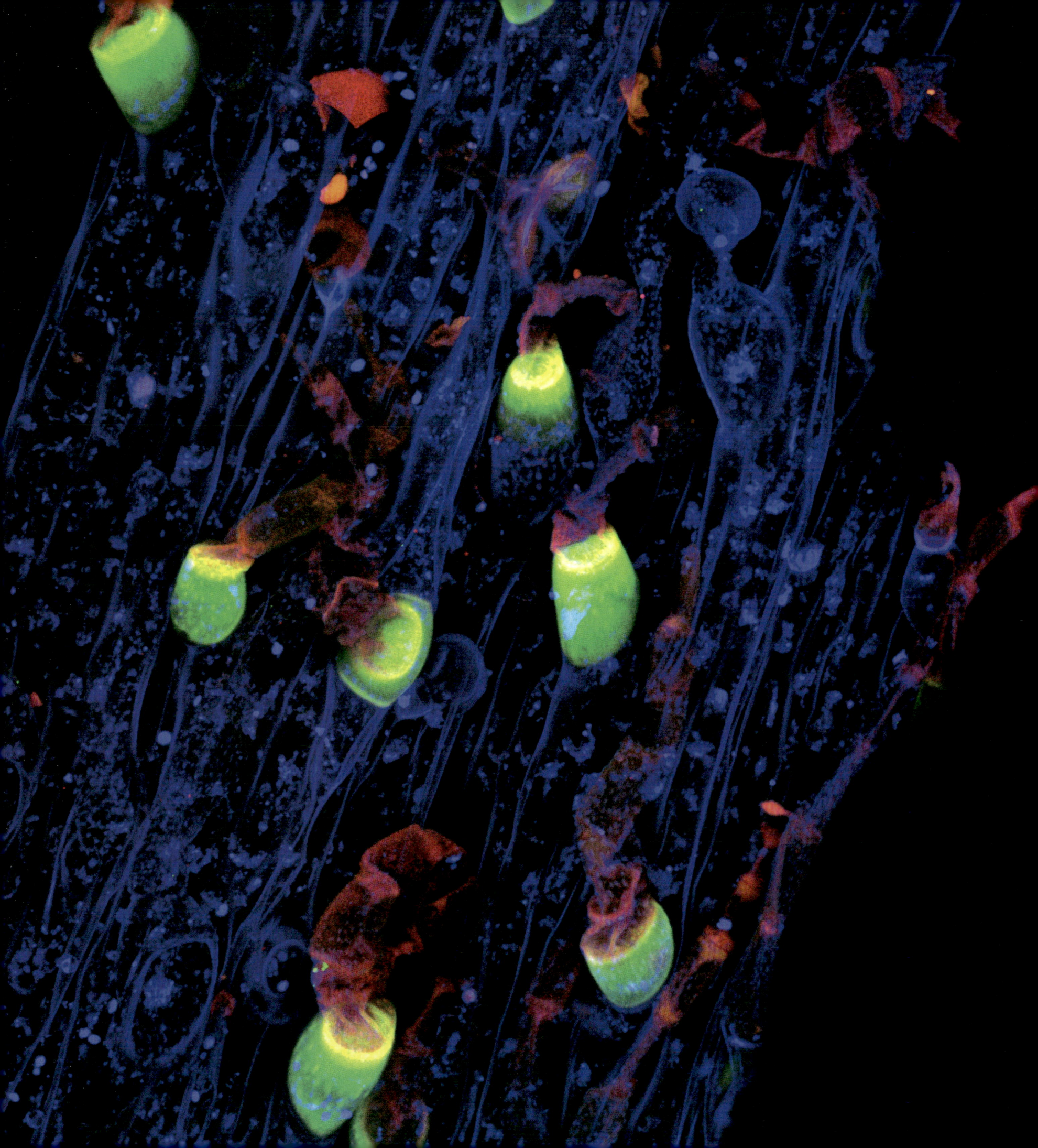

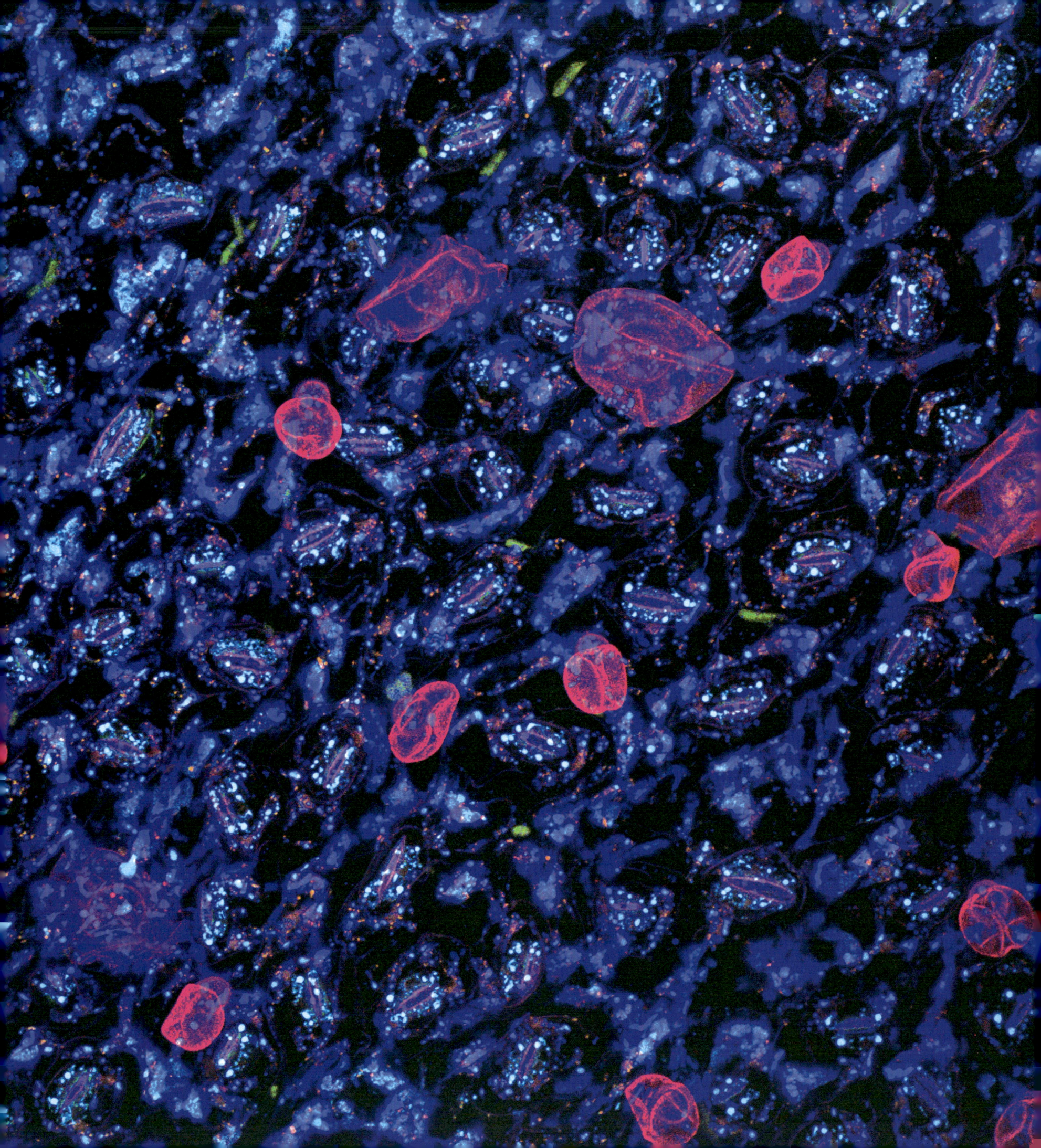

and Clancy Cavnar, which foregrounds psychoactive properties of *Salvia divinorum* in a ritual context conjoined with the knowledge of traditional healers. The author is part of the Human Neuropsychopharmacology Research Group in Barcelona's Hospital de la Santa Creu y Sant Pau, and her study, based on her fieldwork while living in the Mazatec community, opens with a superb natural history of *Salvia divinorum*. Maqueda notes that the Mazatec refer to this plant as "ska pastora", *ska* or *xkà* meaning herb or leaf in the Mazatec language. The name, which connotes a Christian influence, may well have lost a more ancient Indigenous nomenclature that may yet be recovered through interviews with elders. Maqueda clarifies that "the first specimens of living and flowering *Salvia divinorum* that came out of Mexico and constitute the common strain of the plant that has spread throughout the world were collected at the Mazatec Sierra by psychiatrist and ecologist Sterling Bunnell, who introduced them to the United States in 1962". The *Salvia divinorum* variety that circulates commercially should, in fact, be known as the "Bunnell variety", not the "Wasson and Hofmann variety", since these two researchers never exported from Mexico the live plants that they collected. In the section of her study called "Traditional Use", Maqueda reports conducting interviews in a Mazatec town with people suffering from a wide variety of ailments, ranging from vaginal diseases to cocaine addiction to bronchitis, who were cured with different applications of fresh *Salvia divinorum* leaves, usually in a ceremonial context. Maqueda says that some Mazatec consider the plant a female doctor or hold that the feminine healing presence is the Virgin Mary, "while others believe her to be the goddess of plants and animals or the soul of Mother Nature itself". What makes Maqueda's study particularly invaluable is precisely its emphasis on *ritual healing*, and how the *chjota chjine xkà* ("the wise person who cures with herbs") maintains a balance that unites the divine and the earthly in a shared existence. In the section on the therapeutic potential of *Salvia divinorum*, Maqueda has a warning for researchers who insist on working within a strictly Western scientific paradigm that ignores Indigenous wisdom keepers: "it is very important to remember that the traditional use of *Salvia divinorum* by the Mazatec to successfully treat a complex and multifaceted problem like addiction is part of a ritual and a much larger, organic, and inclusive worldview than our compartmentalised interventions, and that the properties of this herb cannot be reduced to the pharmacological mechanism of just one isolated component in the form of a pill". Even so, Maqueda maintains that potential applications that could be developed from salvinorin A include safe analgesics without addictive properties, anti-inflammatories, medications to treat different types of cancer, medications for disorders such as schizophrenia and Alzheimer's disease, antidepressants, medications to treat psychostimulant abuse, psychotherapeutic uses, and neuroprotectors. A truly impressive list, to be sure!

Salvia divinorum (diviner's sage); confocal image

50 μm

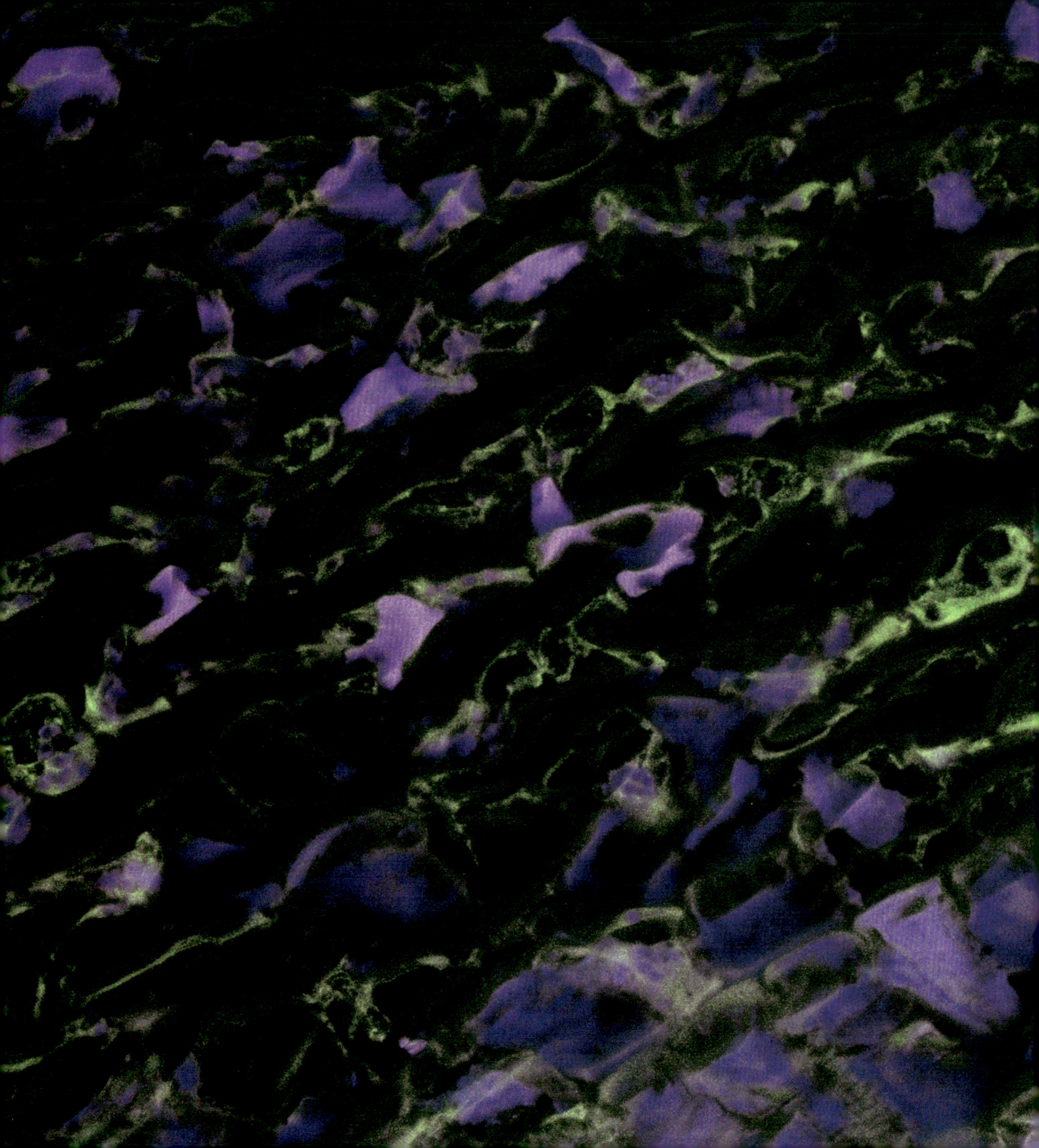

Solanaceae
Solandra maxima
Chalice vine, kiéri, hueipatl

This heavily-branching, fast-growing epiphyte called kiéri by the Wixárika thrives in the wet tropical biome of Mesoamerica and the Caribbean. It has tough oblong leaves and chalice-shaped golden flowers with maroon veined patterns that exude an intoxicating fragrance at night. Traditional Nahua healers in the state of Guerrero, Mexico use this plant for shamanic initiations and divination.

As was the case with so many other plants and fungi in the Americas during the colonial period, Tim Knab maintains that Catholic priests, attempting to prohibit the Huichol (Wixárika) ritual use of *Solandra* (whose common name is kiéri), "probably destroyed many of the plants in their unsuccessful effort to stamp out idolatry in the region".

Masaya Yasumoto says that "the Huichols recognise a close relationship among plants of three solanaceous genera, *Solandra*, *Datura*, and *Brugmansia*". He also points out that "it is believed that the pollen of Kiéri flowers makes birds and insects faint, and causes honey bees to lose their sense of direction". Furthermore, writes Yasumoto, "Kiéri Tewiyari is even more unforgiving, causing madness and even death for the transgressor".

Solandra is a plant of dark mysteries and transformative forces for Indigenous healers willing to assume these considerable risks. Susana Eger Valadez describes the movement between different species, when a human becomes a wolf, under the aegis of this potentially perilous plant-teacher: "the following night, again on the full moon, the wolves who have claimed the initiate will take him into their den. This time, he will be under the influence of the powerful wolf-kiéri plant".

Lilián González Chévez documents the current ritual use of *Solandra guerrerensis* (called hueytlacatzintli) among Nahua healers in the state of Guerrero, Mexico, where

Solandra maxima (chalice vine); confocal image

50 μm

the plant is used primarily to identify the cause of witchcraft and learn how to free victims from these spells, to find lost objects, and to request a specific skill for a client (such as rapidly being able to play a musical instrument with great virtuosity). But, most importantly, *Solandra* still plays an important role in ceremonies of shamanic initiation.

Her superb fieldwork on the contemporary traditional medicinal uses of *Solandra*, also known as hueytlacatl (whose name in Náhuatl means "supreme plant worthy of kings and nobles") includes a detailed description of the complex, days-long curing ritual itself, as well as the first-person testimony of Nahua healer Don Cirilo Soriano from Tlalcozotitlán, a town of 1000 residents in the Mexican state of Guerrero. He gives a compelling account of his powerful plant-visions during his beginnings as a *curandero* decades ago under the tutelage of Teodora Petlatekatl from the town of Zitlala. On the one hand, this story is an intimate tale of resilience recounted by real people describing how plant wisdom from pre-Hispanic times persists to the present day. But it also demonstrates how centuries of prohibition and demonisation of the old gods and the sacred plants themselves have forced ancestral vegetal knowledge underground. Hueytlacatl, once consumed in a cacao drink by Indigenous sovereigns themselves for the purpose of divination, has been recategorised over time as witchcraft, and relegated to a limited, precarious existence in the hands of traditional healers catering to the health needs of a predominantly Náhuatl-speaking population employed by multinational corporations, as Adriana Saldaña Ramírez reveals: these are the poor, marginalised migrant workers exposed to pesticides and contaminated water who struggle to survive as they put fruit and vegetables on the tables of those who live in the global north.

The *santo remedio*, an anthropomorphised plant-god who speaks with the patient in the form of terrifying visions complete with wolves and poisonous creatures, while confirming who caused the patient's disease and by what means, is a powder made from the vine of hueytlacatl and the bark of huaxchiquimolin (*Leucaena matudae*), a member of the Fabaceae family that is an endangered species found only in Mexico. Don Cirilo is quoted as saying that huaxchiquimolin is a "brother" of hueytlacatl and that "they work well together", something that, apparently, he discovered on his own, outside of his apprenticeship. The dose is precisely-measured (1cm deep in a bottlecap). It can then be drunk after dissolving it in water or, preferably, mezcal. To be healed is to have one's shadow restored.

Lilián González Chévez's study ends with a pointed disclaimer in the interest of public safety due to solanaceous neurotoxic tropane alkaloids that can cause insanity and death, and also, one assumes, as a way of calling for the need to protect the sacred Nahua cultural heritage in Guerrero from a tragically destructive outside foreign onslaught of spiritual seekers, similar to what occurred after a publication by Gordon Wasson about another small, remote Mexican town, Huautla de Jiménez, where a healer named María Sabina treated patients with her *niños santos*.

> *"The following night, again on the full moon, the wolves who have claimed the initiate will take him into their den. This time, he will be under the influence of the powerful wolf-kiéri plant."*
>
> ~ Susana Eger Valadez

It has been a privilege and a joy to care for *Solandra maxima* at home in upstate New York. After four years without blooming, hueytlacatl revealed a dozen glorious golden chalices in a spectacular indoor show as a terrible blizzard did its cold work in December. The flowers rose from the vine like green, faceted pyramids, and then extended themselves as pale yellow ghostly forms whose round petalled indentations were eyes that surely saw in different ways than mine. And when they opened, the colour combination of deep gold cups veined in maroon was breathtaking. For weeks at night, each flower was a world unto itself and helped bring a bewitching olfactory landscape from a warmer faraway place that permeated our home. The stamens were thick with sticky white pollen. And it was impossible to resist the temptation to taste it only to become lost in maelstroms of air almost instantly in a dream of being a bee in a dangerous liaison with *Solandra*.

A Mystical Communication with the Kiéri, yarn painting on beeswax and wood,
by Wixárika artist Guadalupe González Ríos (1923-2003)

"A man and his wife have undertaken a pilgrimage to a rocky mountain peak where the tree of wind, called kiéri [Solandra], lives. The plant is depicted in the centre. Its roots spread out like claws on the barren black crag; they do not penetrate the earth. After a long vigil and fast, the devout pilgrims see the flower of the tree bloom at midnight. Under most circumstances its yellow pollen, called tikiyari, is toxic and disorienting, but it fills the cleansed pilgrims with wisdom and magical powers. The gusty wind on the peak carries words from the tree to its devotees (white wavy lines). They cast blood from sacrificed animals as offerings to the tree. They also offer two cornhusks in prayer for a good crop. The white spots are food offerings of corn flour for the spirit of the tree. The flowers are used to sprinkle blood for the deified kiéri."

~ Diana Negrín da Silva, *Grandes maestros del arte Wixárika: acervo Negrín*

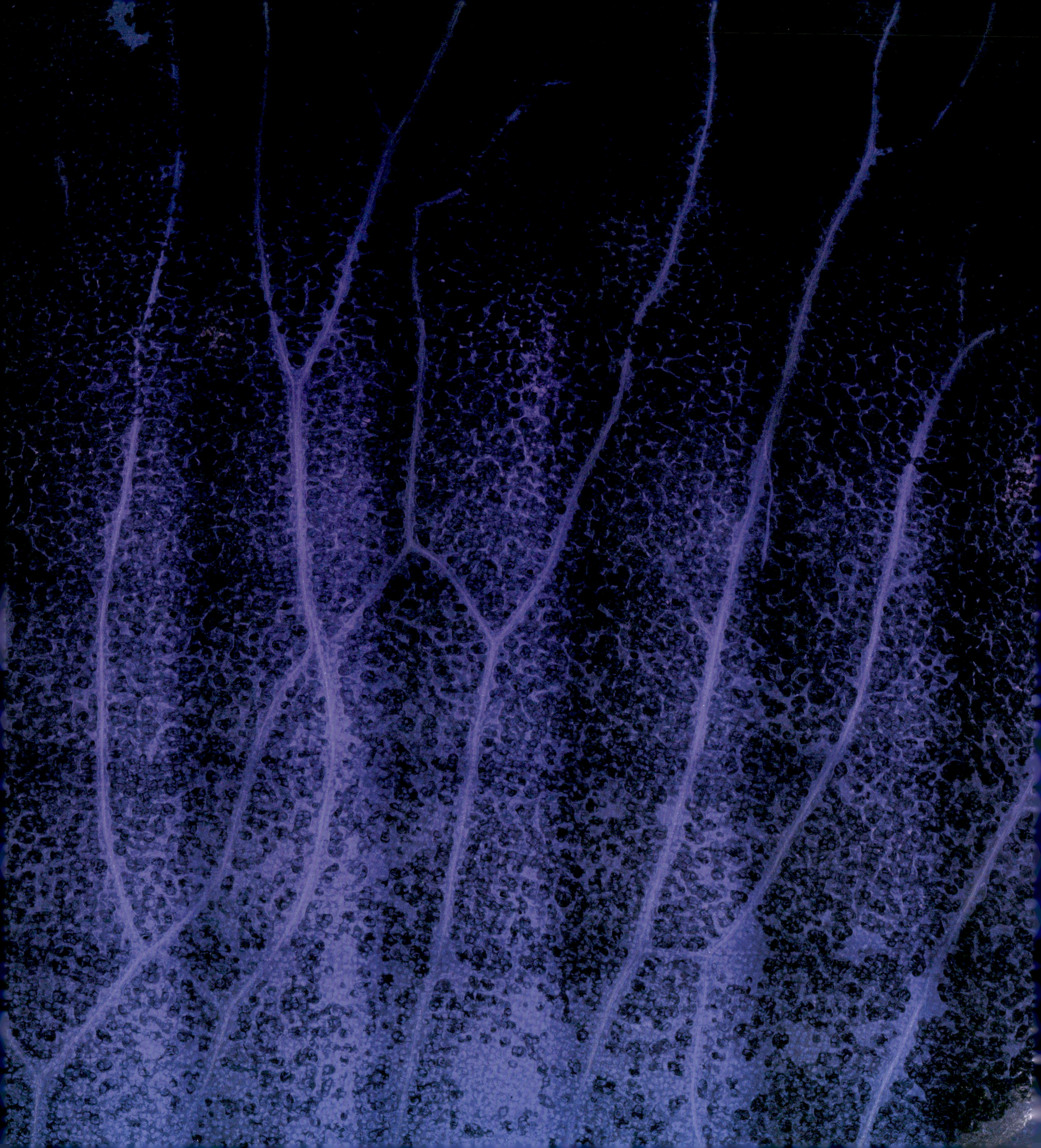

above: *Solandra maxima* (chalice vine); flower
opposite: *Solandra maxima* (chalice vine); confocal image

250 μm

Apocynaceae

Tabernaemontana sananho

Kunakip, sikta, uchu sanango

Tabernaemontana undulata

Becchete, mana heinso

These two closely-related species grow as small trees in the wet tropical biome of the American continent. They have similar tapered leaves and white star-shaped flowers. Their fruit resembles the scrotums of higher mammals. Anthropological and phytochemical research has focused on how these plants are given to dogs to improve their nocturnal hunting skills by sharpening all their senses.

For *Microcosms*, we were able to image *Tabernaemontana undulata*, a species that is very closely-related to the more thoroughly-researched *Tabernaemontana sananho*. Carmen X. Luzuriaga-Quichimbo and her team of researchers from Ecuador and Spain are at the forefront of new studies on *Tabernaemontana sananho*, also known by its common names *kunakip* and *uchu sanango*, as well as its Kichwa name *sikta*. They document medicinal and ritual uses of this small tree that thrives in lowland evergreen rainforests throughout northern South America by the Aguaruna of Peru as well as the Awa, Cofan, Secoya, Shuar, Wao, and Kichwa from Ecuador. The researchers maintain that the Kichwas consider the *Sikta* tree a sacred bridge that "links the person with the hidden forces of Nature", and will not cut these trees down when clearing the forest, keeping their exact location a closely-guarded secret. Different cure-all preparations of Sikta are used as a vulnerary for post-partum bleeding, as a treatment for syphilis, eye wounds, fevers, colds, abscesses, digestive and respiratory conditions, and also as a sedative, painkiller, and contraceptive.

Tabernaemontana undulata (becchete); confocal image

50 μm

According to Christian Rätsch, *Tabernaemontana sananho* is used ceremonially as an admixture to ayahuasca preparations, and to *Virola* snuffs as a "memory plant" that may serve as an aid to remembering the visionary experience more clearly.

Bradley C. Bennett and Rocío Alarcón, in the fascinating article "Hunting and Hallucinogens: The Use of Psychoactive Plants to Improve the Hunting Ability of Dogs", highlight the importance of dogs in Indigenous societies that rely on hunting to survive. They mention how the "Ecuadorian Shuar believe that dogs are a gift from Nunkui, the earth mother", and how, for the Quichua, "dogs are gifts from sachahuarmi or sacharuna (forest spirits)". Bennett and Alarcón consider *Tabernaemontana* and other plant species in relation to ethnoveterinary practices. Citing examples from the Shuar, Quichua and Aguaruna ethnic groups, the authors point out that "in lowland areas of the Neotropics, the primary role of canines is to assist in hunting wild game", asserting that the plants constituting the diet of the canines are meant to improve their prowess as hunters: "plants are employed in baths to reduce their scent or to mask odours and thus decreasing their detectability by the targeted prey. Plants also function to clean buccal and nasal cavities, to enhance olfaction, or to enhance night vision". They bring up the work of Eduardo Kohn, who has studied how the Quichua give their dogs a powerful hallucinogenic preparation called *tsicta* (sikta), consisting of *Tabernaemontana sananho*, in addition to wild tobacco and *Brugmansia* that permits the dogs to "communicate with their masters and to counsel them". Bennett and Alarcón speculate that the improved hunting capacity of the dogs who have been provided this mixture may be due to an efficacious experience of plant-based *synaesthesia* that increases the overall combined enhancement of the ability to smell, see, and hear in what they characterise as a "second sensory or cognitive pathway".

Finally, Bennett and Alarcón treat *Tabernaemontana* as part of a larger vegetal complex by combining "phytochemical data with the ethnobotanical reports of each plant and then classif[ying] each species into a likely pharmacological category: depuratives/deodorant, olfactory sensitiser, ophthalmic, or psychoactive".

above: *Tabernaemontana sananho* (kunakip); flower
opposite: *Tabernaemontana undulata* (becchete); confocal image

50 µm

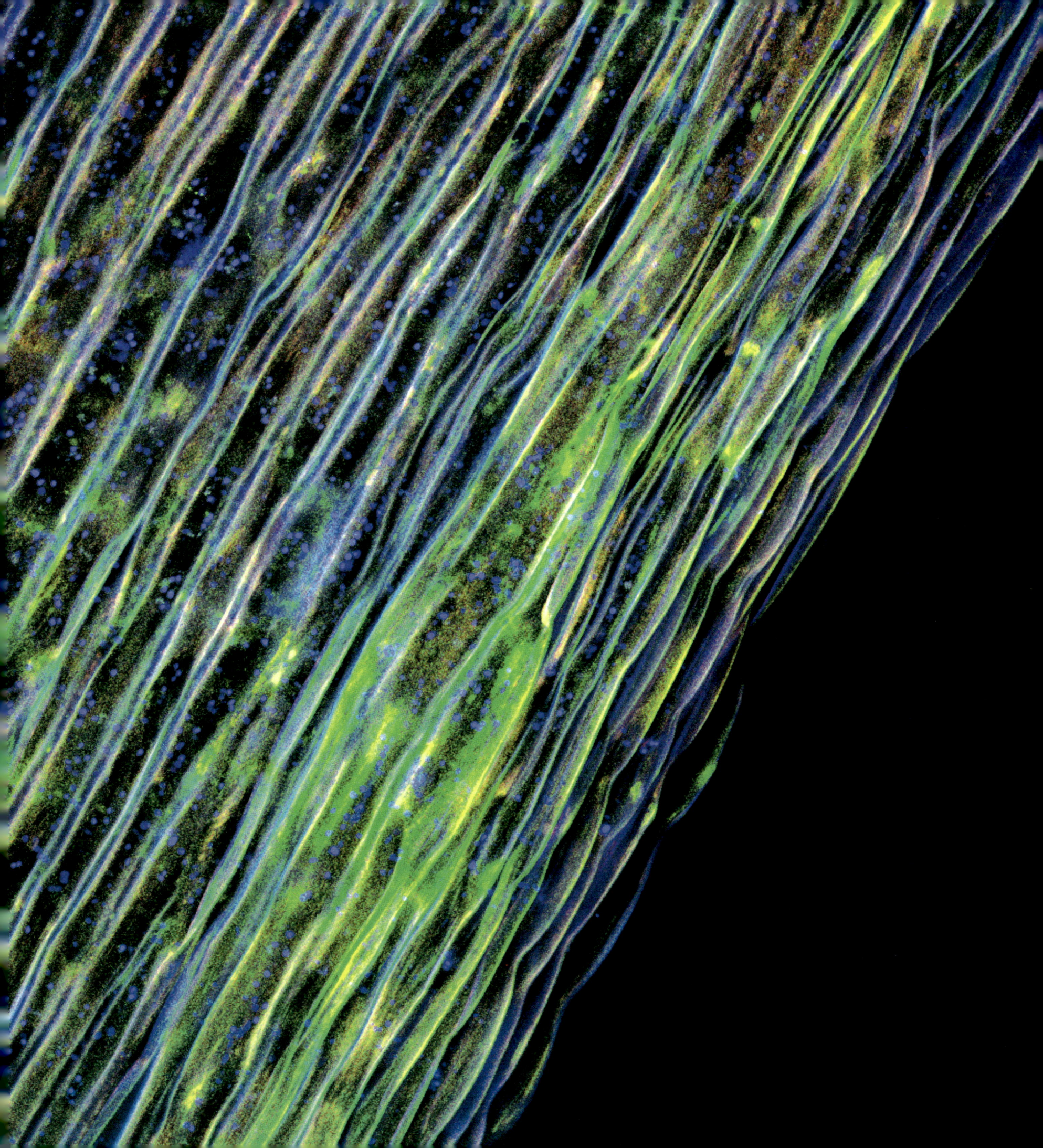

Asteraceae
Tagetes lucida

Cempoal, flor de los muertos, Mexican marigold, Mexican tarragon, pericón, yauhtli

This perennial plant that prefers the subtropical biome has yellow or orange flowers which are associated with the Day of the Dead religious rituals in Mexico. New scientific studies confirm traditional medicinal uses of this plant as a way to treat gastrointestinal pain and rheumatism.

The Huichol (Wixárika) call *Tagetes lucida* (Mexican marigold), a plant native to the Americas, tamutsáli and yahutli. These brightly coloured flowers are used on home altars throughout Mexico during ceremonies associated with the Day of the Dead. Siegel, Collings, and Diaz document how the Huichol mix *Tagetes lucida* with tobacco (*Nicotiana rustica*), which they call yé: "in ceremonial uses, the Huichol smoking of the *yé/tumutsáli* mixture is frequently accompanied by ingestion of peyote, *tesquino* or *nawa* (fermented maize drink) *cái* or *sotól* (cactus distillate), and *tepe* (another alcoholic beverage). Such combinations inevitably produce extremely vivid hallucinations, but less intense visions are obtainable with the smoking mixture alone".

In a fascinating article by a team of researchers led by Laura White Olascoaga, *Tagetes lucida* (which they call pericón) is considered a link that joins the present with a pre-Hispanic past in that yauhtli was strongly connected to the gods of rain and vegetation, and was a symbolic offering as part of this vision of the cosmos. After the Conquest, with the advent of Christianity, *Tagetes lucida* was associated with San Miguel, who was considered the divine protector of harvests, having power over lightning and storms. This belief persists in rural Central Mexico through the use of four bunches of *Tagetes lucida* flowers tied together in the form of a cross and placed over the doorways of farmers' homes and work places.

Tagetes lucida (yauhtli); confocal image

50 μm

Medicinally, the fresh herbage is used to treat abdominal pain, to promote lactation, and to alleviate rheumatism. New research has shown that extractions from *Tagetes lucida* can be used on a large scale as a natural insecticide to protect crops from infestation.

María Eva González-Trujano and a group of Mexican scientists published a study in which "preliminary data provide evidence and give support to the properties attributed to *Tagetes lucida* in the traditional medicine to alleviate pain". The authors highlight this plant's ancient uses among the Nahuatl ethnicity and affirm that *Tagetes lucida* "was and remains one of the most sacred plants used frequently in Mexico". In their conclusion, they affirm that extracts of *Tagetes lucida* "did not produce gastric damage at antinociceptive dose", which is considered "the most common adverse effect of analgesic anti-inflammatory drugs" used in pain therapy. This, therefore, makes *Tagetes lucida* a more appropriate candidate for use in treating gastrointestinal diseases.

Another team of researchers from Mexico headed by González-Trujano and G. Pérez-Ortega, investigated the ethnobotany, phytochemistry, and pharmacology of the tranquilising properties of *Tagetes lucida*. The project was undertaken in Mexico's Morelos State (which was formerly inhabited by Olmecs, Tlahuicas, and Xochimilcas) in recognition of this area's longstanding and "large biocultural heritage that is reflected in the traditional knowledge of the use and management of plants". The research methodology included interviewing eight traditional healers (five women and three men), as well as twenty-five merchants of medicinal plants (seventeen women, eight men), about how they are accustomed to treat anxiety. As a result of these conversations, the scientists discovered that *Tagetes lucida* "was not the most important species for treating ailments of the nervous system, but it was well-known for these properties". In the markets, *Tagetes lucida* was not sold alone, but rather in preparations that were mixtures of "other well-known anxiolytic plants, such as *Ternstroemia sylvatica*, or *Tinantia pringlei*, *Citrus sinensis*, *Passiflora incarnata* or *Agastache mexicana*". In addition, the scientists prepared extracts of *Tagetes lucida* to examine the plant's sedative-like effects on laboratory mice, which did show "a significant but non-dependent decrease in the ambulatory and exploratory activities", which they attributed to the coumarins contained in this plant species used traditionally in Mexican ethnomedicine as a tranquiliser. Scientists from Texas and Mexico led by Julianna Kurpis studied the projected effects of anthropogenic climate change on *Tagetes lucida* in the coming decades and concluded that the habitat of this plant, "a native medicinal plant of important cultural and economic value in Mexico", will become increasingly restricted and fragmented in scenarios modelled for 2050 and 2070. They also maintain that "current suitable habitat is threatened by agriculture, deforestation, and overgrazing". The researchers highlight the importance of studying all these factors so that this information can "be used to help establish sound conservation plans, non-existent to date, for the species". Certainly, this is undeniably prudent ecological advice that might well be applied to a wide range of other plant species, including, of course, those that pertain to *Microcosms*.

Tagetes lucida (yauhtli); flower

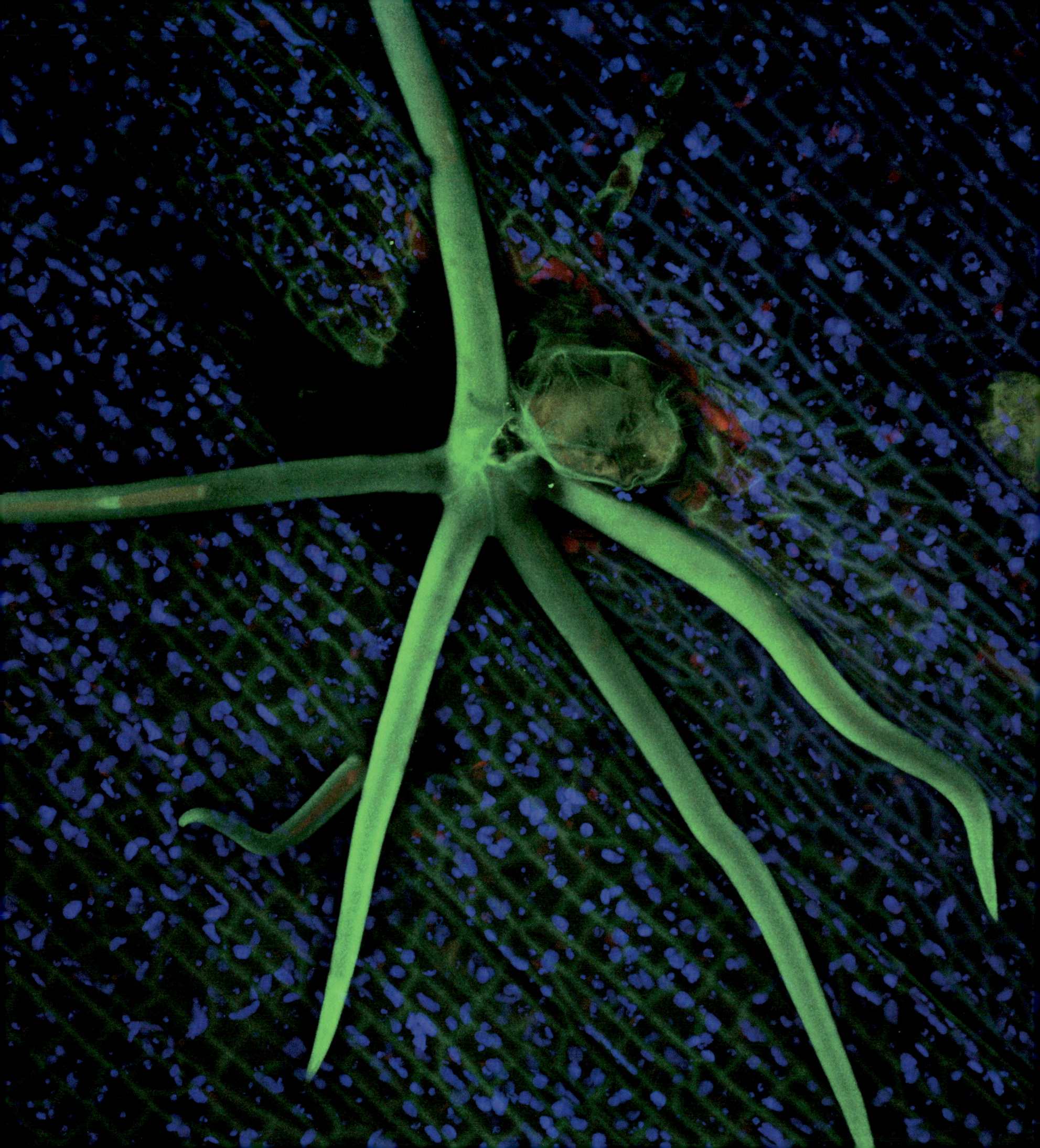

Malvaceae
Theobroma cacao
Cacahuatl, cacao

This evergreen tree with shiny leathery leaves and tiny flowers that grow in clusters directly from the trunk prefers the habitat of wet tropical areas in Central and South America. Its oblong pods contain a white pulp embedded with 20-50 flat seeds, which are used to make chocolate. For the Aztecs, Cacao was a gift from the god Quetzalcoatl, and its seeds were so highly valued that they were used as a form of monetary exchange.

Jonathan Ott is undoubtedly the person who has written most eloquently and profoundly about Cacao. In *The Cacahuatl Eater: Ruminations of an Unabashed Chocolate Addict*, Ott delineates the fascinating history of this plant in Mesoamerica and, in so doing, creates a broader definition of the sacred and why certain plants have more cultural significance than others. According to Ott:

"Mexican tradition holds that the man-god Quetzalcoatl had been led into the lost paradise, wherein dwelt the children of the sun god. When he returned to the world of men, Quetzalcoatl brought with him the seeds of *cacaoquauitl*, our beloved *cacahuatl*. Thus stimulated, he gathered disciples, taught them the civilised arts of agriculture, astronomy and medicine, and became the ruler of Mexico". Quetzalcoatl then cultivated cacao in his garden, nourished himself with its seeds and became inebriated with the liquor made from the fermentation of the pulp of the cacao fruit. Ott mentions another important characteristic of cacao as a medicinal plant and admixture: "like ayahuasca, *cacáhuatl* was an all-purpose pharmaceutical vehicle for administration of many medicinal plants; both curative specifics and shamanic inebriants".

In *The Falling Sky*, Yanomami shaman and social activist Davi Kopenawa tells the story of how in a remote past his ancestors were crushed or thrown underground except

Theobroma cacao (cacao): confocal image

50 μm

in one place where the sky finally came to rest on a wild cacao tree, which bent under its weight but did not break. The first people were then able to escape through a hole created by this tree's canopy.

In a chapter on cacao that appeared in *Fruit and Vegetable Phytochemicals: Chemistry and Human Health*, a team of Mexican researchers with lead writer Alfonso A. Gardea state that "cacao was first presented to Charles V in Spain as a chemical weapon, based on the belief that once warriors had taken the drink, they were capable of fighting nonstop for the whole day. However, the sour and even pungent cocoa drink – prized by the Aztecs – was not wholly acceptable to local taste". The authors go on to say that, globally, "cocoa production has given rise to serious concerns about sustainability because of factors such as child labor, harsh working conditions, abuse of the environment, changing weather, and low profit". Did you know that the largest chocolate company in North America, Hershey's, generates US$3.72 billion in annual sales and that most of the cocoa beans they purchase are from Ivory Coast, by far the world's largest producer at 1.4 million tons annually? In terms of cocoa's biologically active components, studies have shown that cocoa lowers blood pressure and also prevents cardiovascular disease. It also "has been reported to increase the total antioxidant capacity in human blood plasma".

A 2023 study by a team of scientists from the Borough of Manhattan Community College (City University of New York) and Rutgers University led by Nadjet Cornejal investigated the antimicrobial and antioxidant properties of *Theobroma cacao*, which, as the authors mention in their introduction, originally was domesticated in the upper Amazon region some 1500 years before these Indigenous groups from South America migrated with this plant to Mesoamerica. The authors cite research demonstrating how cacao beans "were also used for currency, for trade, for ritualistic practices, and for large feasts among Mayans and among Aztec elites". There is also abundant and aesthetically-appealing archaeological evidence depicting cacao in jade, obsidian, stonework, and pottery. The comparative study of four plants (*Theobroma cacao*, *Bourreria huanita*, *Eriobotrya japonica* and *Elettaria cardamomum*) widely used in traditional medicine in Central America showed that whole cacao beans from La Antigua, Guatemala "showed the highest total phenolic concentration, antioxidant activity, and selective antiviral activity".

Three researchers from the University of California, Santa Barbara, led by Anabel Ford published a study in 2022 that no doubt will upend widely-held conceptions that cacao in ancient Mesoamerica was reserved for use by an elite. They analysed 54 sherds (broken, jagged pieces of pottery from an archaeological site) from Late Classic Period Maya residential contexts around El Pilar (Belize/Guatemala). Using the technique of laser mass spectrometry, they detected "a significant amount of the key biomarker of theophylline, to signify cacao". Using this cutting-edge technology to undertake chemical residue analyses, the authors hoped to answer questions such as: "what is cacao consumption among the Maya populace? Is consumption restricted to higher ranking houses? Are farmers who might grow cacao, like those of Ceren also consuming cacao?". Based on the findings of this fascinating study, the scientists affirm: "these results dispel any doubt as to the importance and inclusiveness of cacao consumption among the Late Classic Maya. That cacao is generally available does not diminish its value but contextualises its formal and ceremonial importance as a cultural phenomenon that experienced wide participation by the populace. Well beyond the elite ritual civic-ceremonial realm, we interpret the identification of cacao in vessels belonging to people of all walks of life as confirmation that cacao's prestige was consumed by all in Maya society".

opposite top left: *Theobroma cacao* (cacao); open pod with cacao "beans"
opposite top right: *Theobroma cacao* (cacao); Finca Santa Ana, Pueblo Redondo, Telica, León, Nicaragua
opposite bottom: *Theobroma cacao* (cacao); with its fruit growing directly from the trunk

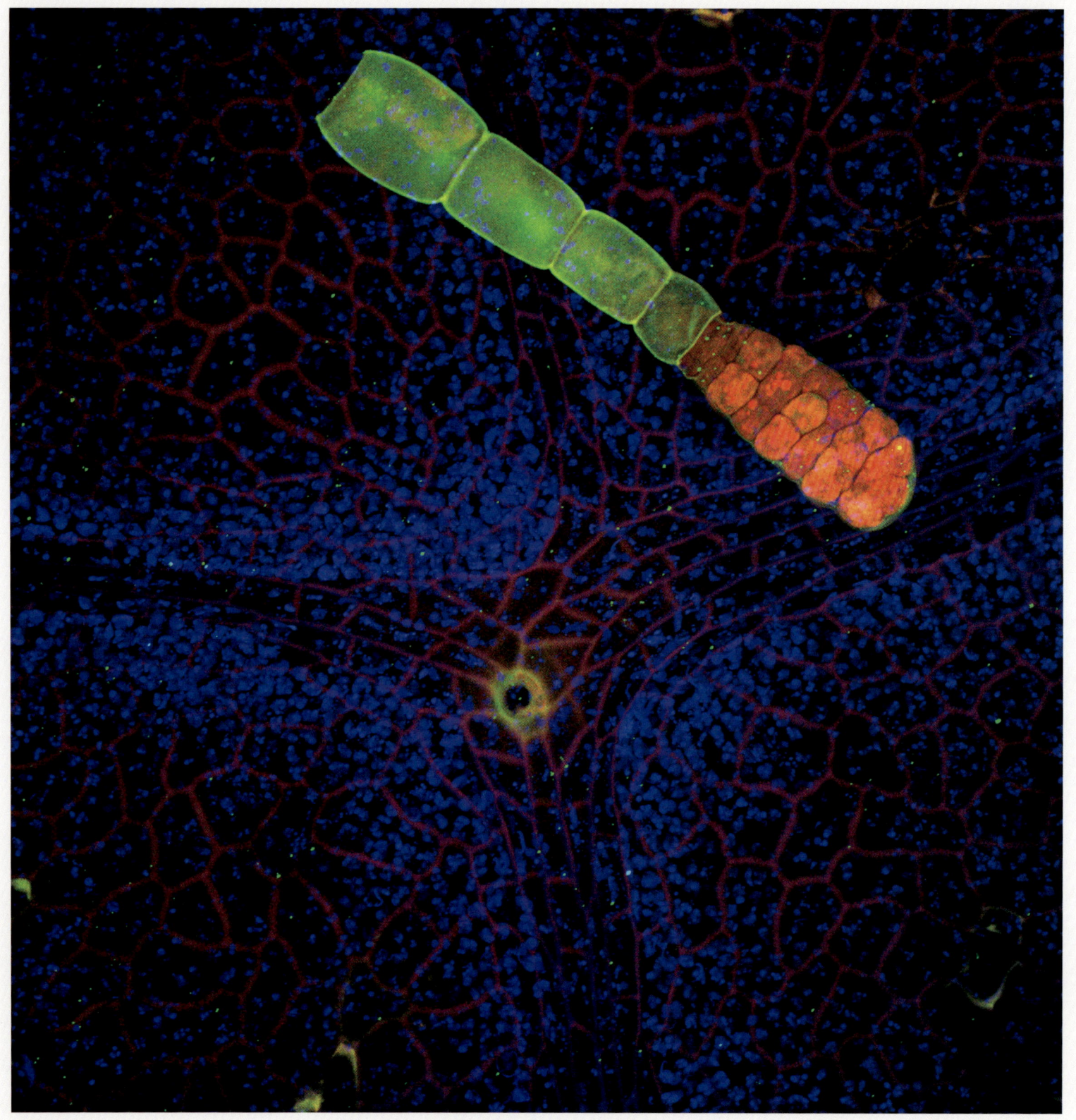

above: *Theobroma cacao* (cacao); confocal image

50 μm

opposite: *Theobroma cacao* (cacao); confocal image

50 μm

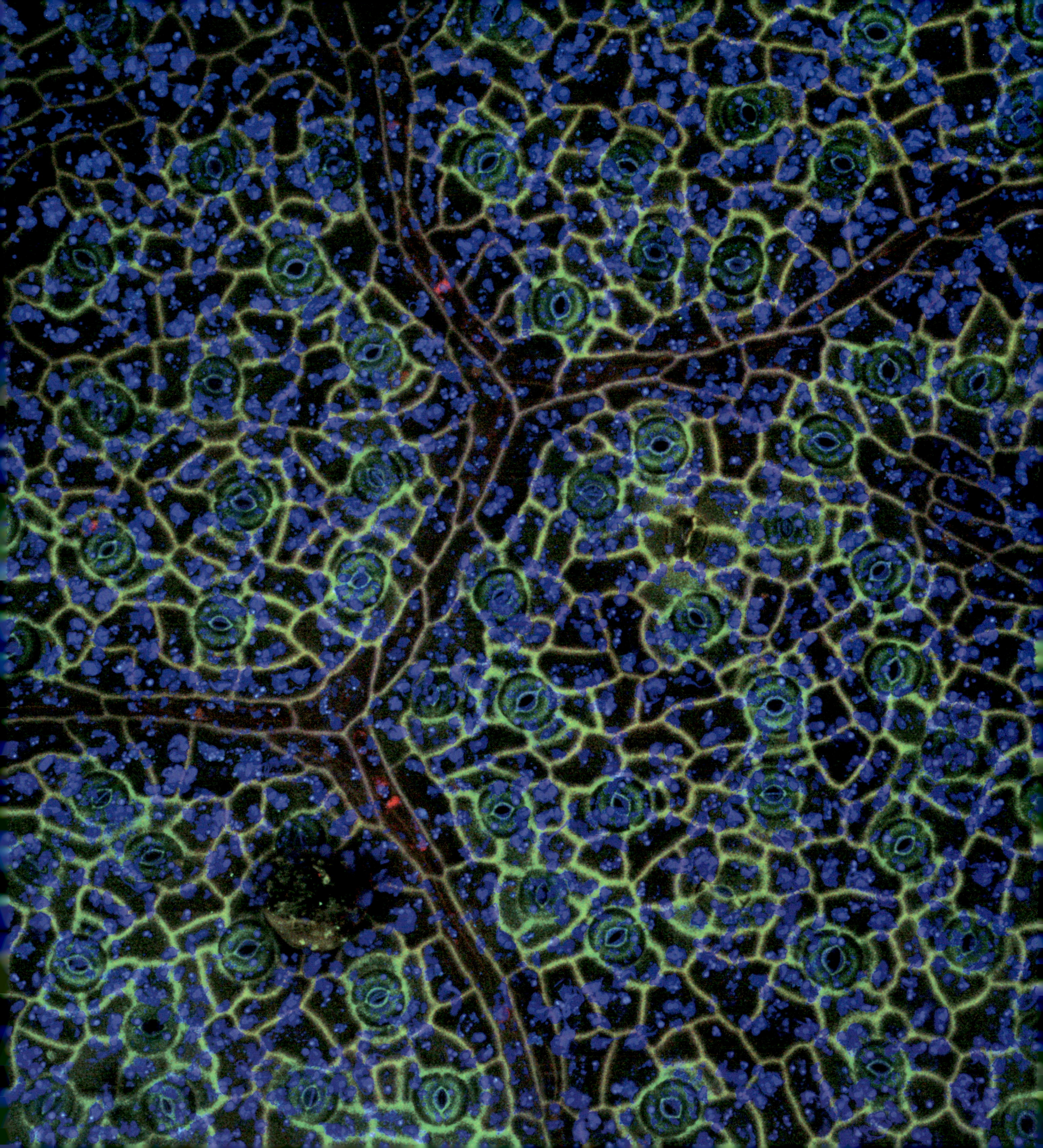

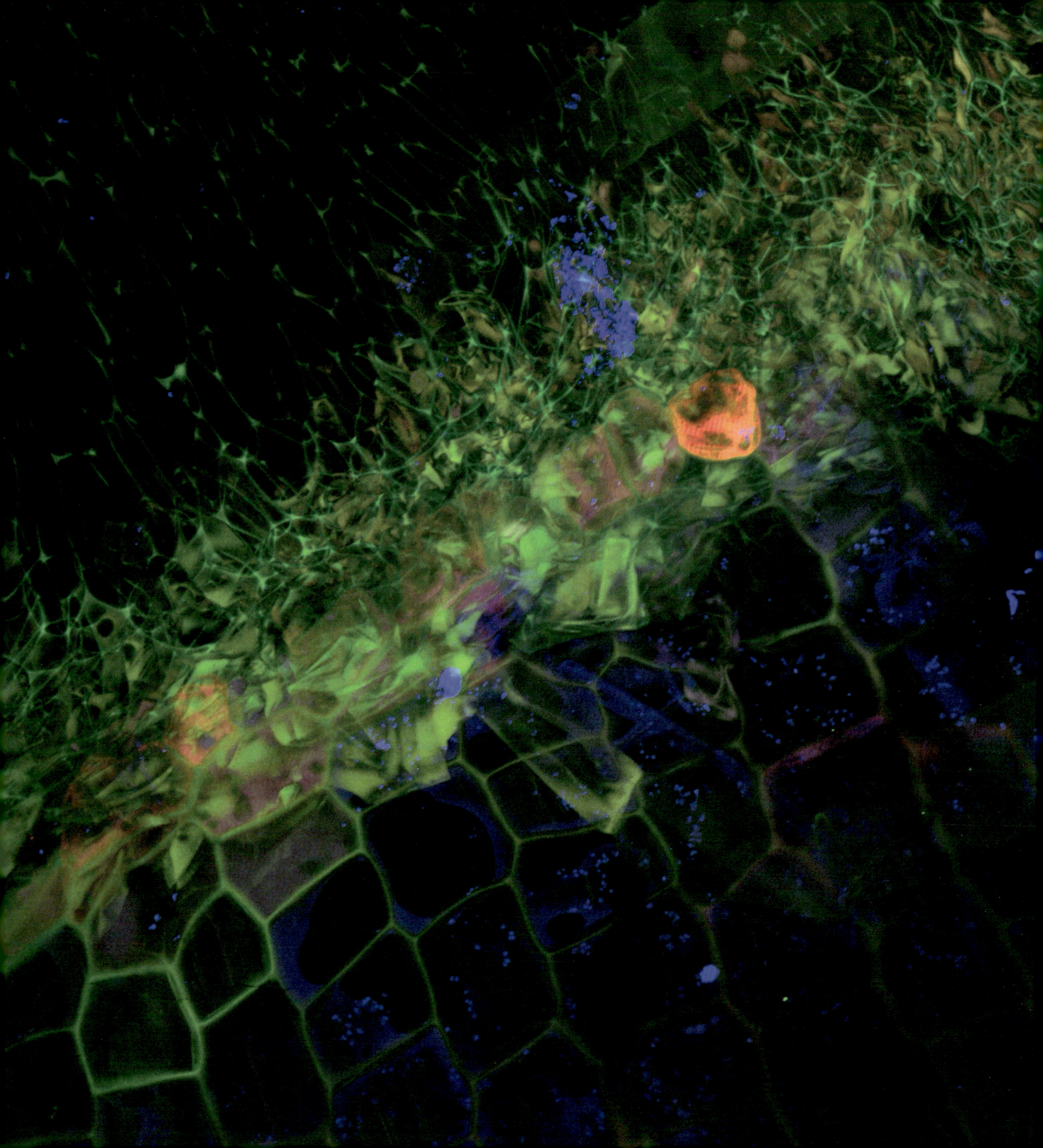

Cactaceae
Trichocereus macrogonus var. pachanoi

Aguacolla, cactus of the four winds, huachuma, lapitoj, San Pedro cactus

Traditional use of this dark green columnar cactus that contains mescaline is particularly well-conserved in northern Peru. In the pre-Hispanic era, especially in the area of Chavín de Huantar, the Huachuma cactus was the basis for a shared ritual exchange between different ethnic groups. There is great concern now about how the habitat destruction of mining projects and unsustainable harvesting for recreational purposes have seriously endangered the San Pedro cactus.

Perhaps the most compelling new research on the ritual use of *Huachuma*, the San Pedro cactus, is by Argentine Leonardo Feldman. He points out that San Pedro is one of the *plants of power* that is best represented in pre-Incan iconography, appearing in the art of a variety of Indigenous cultures such as that of Chavín (with its utterly riveting anthropomorphic feline Bearer of San Pedro), Nazca, Moche, Paracas, and Chimú. For Feldman, it was exhilarating to see the cactus growing freely and abundantly among the ruins of the ritual centre of the Templo del Lanzón in Chavín de Huántar.

The traditional use of this cactus, which contains significant amounts of mescaline, extends to northern Chile, northwestern Argentina, Bolivia, Peru, and Ecuador. Even so, he says, the pre-Hispanic uses of the cactus (as a sacrament that facilitated communing with the divine spirits of nature) are still particularly well-conserved in northern Peru in the mountains of Piura (Huancabamba and Ayabaca), where the Complex of Las Huaringas Lagoons (a UNESCO World Heritage Centre) with its *páramo* ecosystem is located.

Trichocereus macrogonus var. *pachanoi* (huachuma / San Pedro cactus); confocal image

50 μm

left: Feline deity bearing Huachuma, Chavín de Huantar, Peru

right: *Trichocereus macrogonus* var. *pachanoi* (huachuma / San Pedro cactus) near the sacred ceremonial site of Chavín de Huantar, Peru

Feldman analyses the social function of San Pedro as a means of diagnosing illness and healing as well as conflict resolution or achieving prosperity in diverse forms. It is also employed in rituals for predicting the weather, doing astronomical observation, and extracting the *mamayacu*, (the mother of the sacred waters of the lakes).

Even in the present day, Feldman affirms, the traditional use of San Pedro "represents a factor of social cohesion and regional cultural identity, while at the same time preserving a centuries-old religious system". In the pre-Hispanic past, the Huachuma cactus may well have served as the basis for a pan-Andean religious and political *lingua franca* that enabled people from different ethnic groups to communicate, to mediate their differences and to coexist by means of a shared ritual exchange.

A team of academic researchers in Media, Social Anthropology, and Journalism from Ecuador and Spain led by Ángel Torres-Toukoumidis investigated the ancestral healing rituals associated with the cactus aguacolla (*Trichocereus macrogonus* var. *pachanoi*) in relation to "touristic, historical, and patrimonial repercussions" in rituals conducted in the community of Ilincho in the Andes of southern Ecuador. The ceremonies are led by members of the Saraguro ethnic group, who speak Runashimi (a Kichwa dialect) as well as Spanish.

For the researchers, "the audiovisual medium allows those brief moments of intimacy and recollection to be captured" between healers and visitors in terms of "cognitive processes that are not verbal". The ceremony filmed by the researchers at the Health and New Life Foundation at the Yachak Center in Ilincho was conducted by Yachak Polibio Japón and lasted approximately ten hours. Yachak is a Kichwa word meaning "wise". The researchers distinguish between four categories of medicinal knowledge used by the Saraguros: 1) wachakhampiYachak (midwives who work with pregnant women and babies), 2) yurakhampiYachak (someone who uses plants to cure diseases such as headaches or fevers), 3) kakuyampiYachak (a person who treats bone and joint problems) and 4) rikuyhampiYachak (a healer "who uses entheogenic plants to cure supernatural diseases in night sessions called *mesadas*)". In their conclusion,

the authors, state that the "aguacolla is the main element in the mesada [altar]" and resembles "a central cosmic tree that commands the place of the ceremony".

A series of fourteen lagoons in a zone of stark beauty in Northern Peru called Las Huaringas is the setting for healing ceremonies based on the ancestral knowledge of sacred plants, including the cactus *Trichocereus macrogonus* var. *pachanoi* (huachuma). A team of researchers led by Peruvian Miguel Ruiz published a 2024 book chapter that examines the parameters of what they characterise as an increasingly popular "mystical tourism" industry in this high Andean region of Piura near Huancabamba. Their goals included interviewing shamans and tourists in order to present a fuller comprehension of how these services benefit national and international visitors with a wide variety of ailments and desire for spiritual experiences as well as the area's economy. They also examine the perspective of local residents, especially those who are concerned with "the distortion of shamanic practice with the presence of fraudulent healers who take advantage of tourists' needs". Their study, conducted in December of 2023, focused on Laguna Negra and Laguna Shimbe and, in terms of methodology, "adopted a phenomenological design because it aimed to collect information during or shortly after participants completed ritual experiences". The researchers found that participants did indeed experience "cultural authenticity, which, for the residents, means a cultural heritage passed down from generation to generation". The efficacy of the shaman, who invokes in the ceremonies both a Christian God and ancestral Incan deities, depends on the trust and faith of the participants. The magnitude and natural beauty of the landscape also "highlight the mystical environment of the area", making it an ideal place for the "blooming baths" and healing rituals with "potions that induced altered states of consciousness". In terms of planning for future research, the authors point out that "some shamans may seek ways to integrate their practices with Western medicine or spiritual tourism, while others may choose to maintain a more traditional approach and resist external influence".

In her article, "Save a Dragon, Slay the Grail", from *How Psychedelics Can Help Save the World: Visionary and Indigenous Voices Speak Out*, a collection edited by Stephen Gray, Laurel Sugden sounds the alarm about the forces that threaten *Trichocereus macrogonus* var. *pachanoi* in its Andean countries of origin, especially Peru. These include habitat destruction through mining and construction projects but, most significantly, unsustainable harvesting practices that produce cacti destined for San Pedro tourism in Cusco, and for illegal export from Lima as San Pedro powder and chips. Sugden, along with her husband, the maestro huachumero Josip Orlovac del Río, co-founded the Huachuma Collective, which has created an alliance of healers, Indigenous leaders, and Andean community members in order to "protect, conserve, and plant Huachuma, and explore sustainable practices for growing and working with traditional medicine in Peru".

Basellaceae
Ullucus tuberosus
(aborigineus)

Ancestral potato, melloco, olluko, Papa lisa

This subspecies of the potato is cultivated for its edible tubers and nutritious spinach-like leaves in the higher Andes, and is likely the forebear of the domesticated varieties. The tubers have a brightly-coloured waxy skin in a striking range of yellows, pinks and purples. According to Quechua narratives about Andean food plants, the potato (Mama acxo) was revered as a mother. Indigenous families continue practicing traditional agricultural techniques and preserving the genetic diversity of almost 1500 varieties of potatoes.

The Peruvian historian María Rostworowski de Diez Canseco explains and recounts a mythic narrative related to Andean food plants, including, of course, the potato, which was domesticated in Peru 7000 years ago: "the feminine and divine element represents the fruitful and prolific mother; not in vain was the earth called Pachamama (mother earth) in the Quechua language, the sea Mamacocha, the moon Mama Quilla and also all the plants useful to humans were called and adored by the name Mama (mother): Mama sara (maize), Mama acxo (potato), Mama oca (oca, a native Andean tuber), Mama coca (coca shrubs). An example of the cult to femininity, and the woman who fills her children with goods, is the myth of the goddess Raiguana. Natives tell that in ancient times humans had nothing to eat and, in order to obtain food, they asked Yucyuc for help. Yucyuc was a little bird with a yellow beak and feet, smart enough to obtain the seeds of staple crops kept by Mama Raiguana. To achieve this, Yucyuc requested from Sacracha (another bird) a handful of fleas which he threw in the eyes of the goddess. Raiguana sobbed her eyes out and for a moment lost her son called Conopa. An eagle caught the child from his mother's arms. Raiguana had to promise to share the seeds with the humans if she wanted to have her son back.

A variety of Andean potatoes from the market in Arequipa, Peru

To the people of the highlands she gave potatoes, oca, olluco, mashua (native tubers), and quinua (native grain), while the coastal inhabitants received maize, cassava, sweet potatoes and beans".

Conopa is not only the name of the son of Raiguana, the goddess who is the guardian of all food plants in this mythic tale. It is also the word for the protective spirit of each crop, the best part of which served as a ceremonial offering and tribute to the gods to ensure maximum yields in the future.

The Inca religion was thoroughly linked to the cycles of the successful production of food, despite droughts, plagues, and cold snaps. Their gods were protagonists in this process and Nature itself was deified. In the divine hierarchy, there were major and minor gods as well as goddesses of the terrestrial world associated with a maternal, fertile earth and the plants consumed as food. As the myth suggests, each of these plants, including for example, the potato (Mama acxo), is revered as a mother.

Ben Kamm, the US ethnobotanist and founder of Sacred Succulents, collected the *Ullucus tuberosus (aborigineus)* from an ancient Incan agricultural terrace near Cuzco that Jill Pflugheber and I were honoured to be able to image with the confocal microscope. According to Ben, it may very well be possible to consider this plant the wild ancestor of many of the varieties of potatoes that over time were domesticated, conserved and consumed by the Andean Indigenous population in a region with a wide array of climates (some extremely harsh) that required great diversity of plant ecotypes.

Here is how Ben describes the trip to Peru when he found this plant:

In 2010 we made our second visit to the Incan outpost of Pumamarca situated at about 12,000ft [3650m] on a spur jutting out above the Patacancha Valley. From there we made the spectacular hike back towards the Inca's final holdout in the Vilcanota Valley: Ollantaytambo. Revelling in the jubilant display of wildflowers and regrowth of native Alnus, Escallonia, and Myrcianthes trees, our path took us along the most heavily terraced mountainside I have seen in all of the Andes. With over a thousand stone terraces it would have been a site of incredibly intensive agriculture.

There are some moments that conspire towards the sublime – the angle of the sun diffusing through a lacy wisp of cloud; the exhalation of moist earth, sunbaked stone, vegetation and wildflower combine to perfume just so, and your quid nestled comfortably between gum and cheek; the perfect combination of leaf to llipta to saliva infuses the world with an undeniable grace.

Thus it was about halfway to Munaypata, near 10,500ft [3,200m], that I noticed some long trailing stems hanging down the rough stone terrace walls. Closer inspection revealed a lovely pink hue to the stems and semi-succulent leaves that looked and tasted very similar to the cultivated Ullucus. Following one vine along its route of growth I discovered a small spire of mini-star flowers affirming the plant's identity. I also observed some odd thread-like stems shooting off from a few leaf nodes, these disappeared into the cracks of the terrace wall. I was able to locate one that terminated in a relatively large dirt-filled fissure and with careful excavation uncovered several small pearlescent-pink tubers!

This is considered the wild form or ancestor of the Andean staple crop "ulluco" ("Papa lisa"). Cultivated ulluco very rarely sets seed and it is possible that this wild subspecies, which seeds more readily, could be used in breeding programs. It has also been speculated that it was used in breeding new varietals by the Incas. It is plausible that what we discovered was an anthropogenic relic … or it could just be this wild subspecies, which we've since observed as a cliff dweller, that favored the rocky habitat of the terracing.

According to a study by Tapia and de la Torre, "potato is the prototype crop of the Suni agro-ecological zone together with the Andean tubers oca, olluco and mashua". They go on to say that many Indigenous families continue traditional agricultural practices and have "preserved and added to crop genetic diversity". At the Potato Park near Cuzco, there are 1367 varieties of this plant that, in the future, may be a key to staving off worldwide hunger.

Tapia and de la Torre document the quintessential importance of Andean "women's participation in plant genetic resource conservation", knowledge that these women transmit from one generation to the next.

A group of scientists from Medellín, Colombia led by Nathalie Heil investigated the wound healing properties of aqueous extracts of *Ullucus tuberosus*. They found "an increase in collagenase activity of 12%" which makes *Ullucus tuberosus* "a promising candidate to support scarless tissue regeneration".

top left: Incan terracing from Punamarca to Munaypata, Cusco, Peru
top right: *Ullucus tuberosus* (*aborigineus*) (ancestral potato); from Punamarca to Munaypata, Cusco, Peru
above: Peruvian native potatoes harvested in Cusco, Peru

Ullucus tuberosus

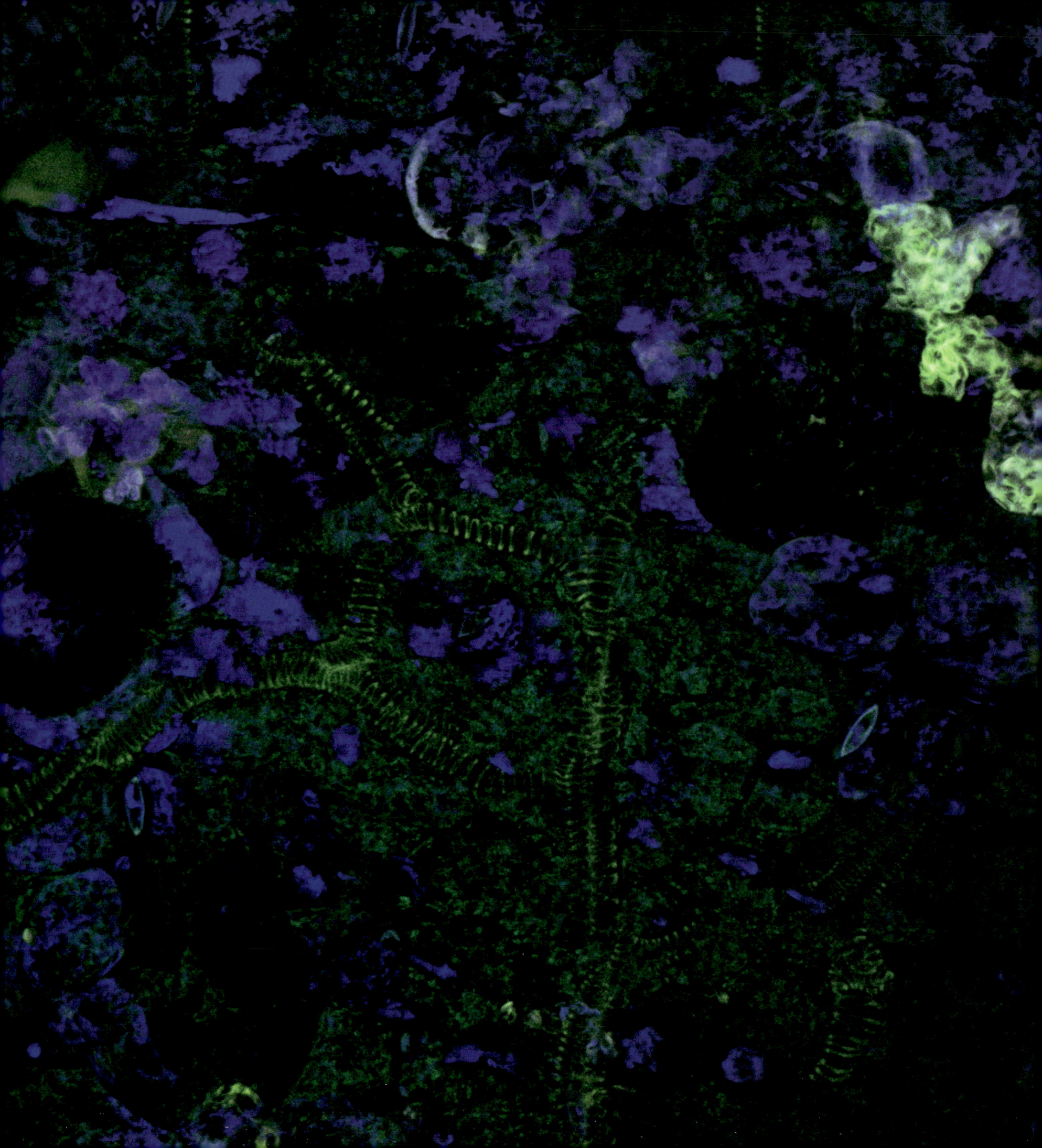

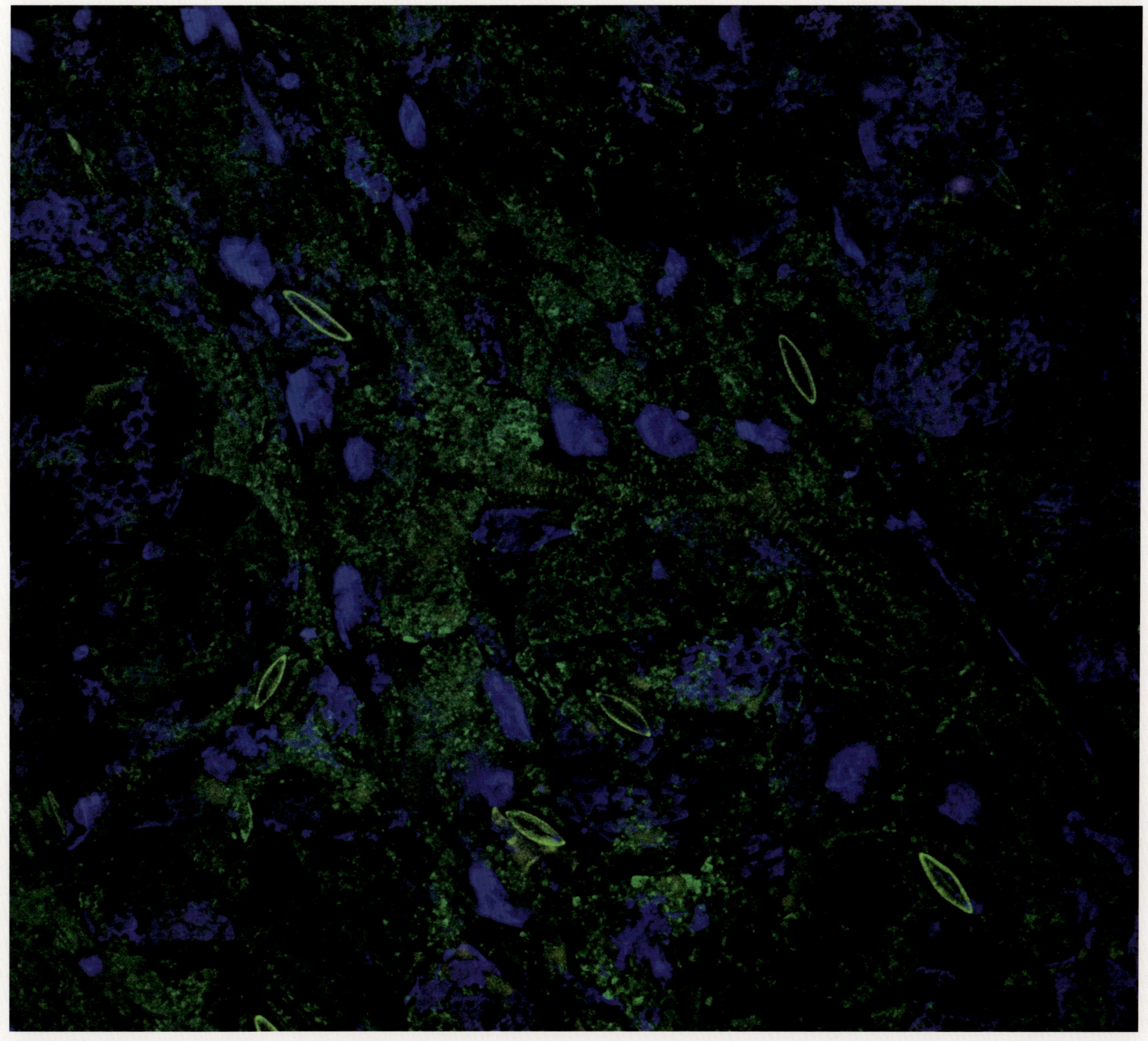

above: *Ullucus tuberosus* (*aborigineus*) (ancestral potato); confocal image

50 μm

opposite: *Ullucus tuberosus* (*aborigineus*) (ancestral potato); confocal image

50 μm

Myristicaceae
Virola theiodora
Cumala, epená, yãkoana

This slender tree in the nutmeg family from the rainforests of the western Amazon basin produces a red "resin" in the inner bark that is used by Indigenous groups to make yakoãna. This powerful psychoactive snuff facilitates communication with the spirit world, and helps traditional healers such as Yanomami social activist Davi Kopenawa protect the territorial rainforest from the destruction of outsiders.

Indigenous healer and spokesperson Davi Kopenawa collaborated for decades with French anthropologist Bruce Albert to produce *The Falling Sky: Words of a Yanomami Shaman*, an amazing and inspiring life story that Albert calls a "cosmoecological prophecy about the death of shamans and the end of humanity".

The book recounts, in excruciating detail, the struggle of the Yanomami to save their land and cultural traditions from the incursions of missionaries, road builders, government workers, gold prospectors, and cattle ranchers.

These conflicts have become far more violent, frequent, and potentially apocalyptic under the racist Brazilian government of Jair Bolsonaro, who recently said, "Indians are undoubtedly changing… They are increasingly becoming human beings just like us".

Kopenawa, as a young man, decided he needed to learn Portuguese, saying, "isn't it up to me to defend our forest?".

In *The Falling Sky*, he declares: "we do not want our forest to die, covered in wounds and the white people's waste. We are angry when they burn its trees, tear up its floor, and soil its rivers…The paper skins of their money will never be numerous enough to compensate for the value of its burned trees, its desiccated ground, and its dirty waters".

Virola theiodora (yãkoana); confocal image

50 µm

His insights and social activism are enhanced through his lifelong relationship with yãkoana (*Virola theiodora*, from the Nutmeg family), a powerful visionary snuff made from the resin of the great tree's bark.

He says, "I defend the forest because I know it thanks to the power of *yãkoana*".

He goes on to say: "if we do not feed the spirits with the *yãkoana*, they sleep in silence and our thought remains closed".

A team of Brazilian scientists headed by Inês Ribeiro Machado published an overview in 2021 of *Virola surinamensis*, a closely-related species to *Virola theiodora* in which they summarise research related to the plant's medical, therapeutic and pharmacological properties as antimicrobial, larvicidal, antitumor, antinociceptive and anti-inflammatory, antioxidant, leishmanicidal, antimalarial, cercaricidal, trypanocidal, gastroprotective, and antiulcerogenic. In a section on traditional medicinal uses of *Virola*, the authors mention that the Waiãpi Indians of Amapá inhale the vapour of the plant's leaves as a treatment for malaria. *Virola surinamensis* is also used as an insect repellent, for the healing of wounds and also for "inflammation of the digestive, urinary, and reproductive systems". Known as *ucuuba*, *Virola surinamensis*'s seeds and bark are used by different ethnic groups as hallucinogens in their rituals. Ucuuba seeds are also of interest in the cosmetics industry.

top: Yanomami psychoactive snuff preparation using *Virola* tree bark

above: *Virola calophylla* (yãkoana); seed with its red-orange aril

opposite: *Virola calophylla* (yãkoana); in the upper canopy, Sucusari, Peru

overleaf left: *Virola theiodora* (yãkoana); confocal image

50 μm

overleaf right: Davi Kopenawa, Yanomami activist, traditional healer, author of *The Falling Sky* and cast member of *The Last Forest*, Cannes Film Festival, 2024

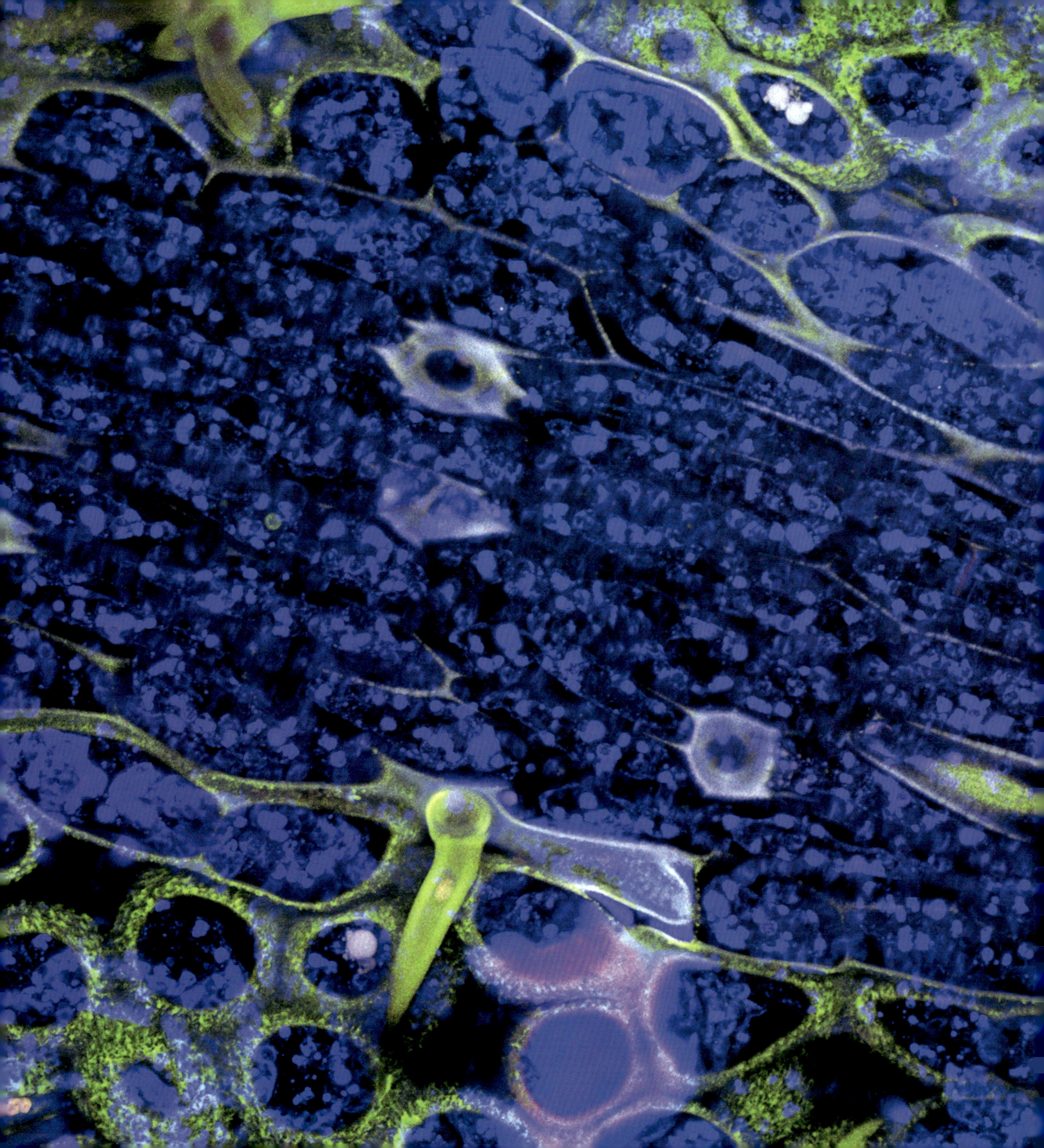

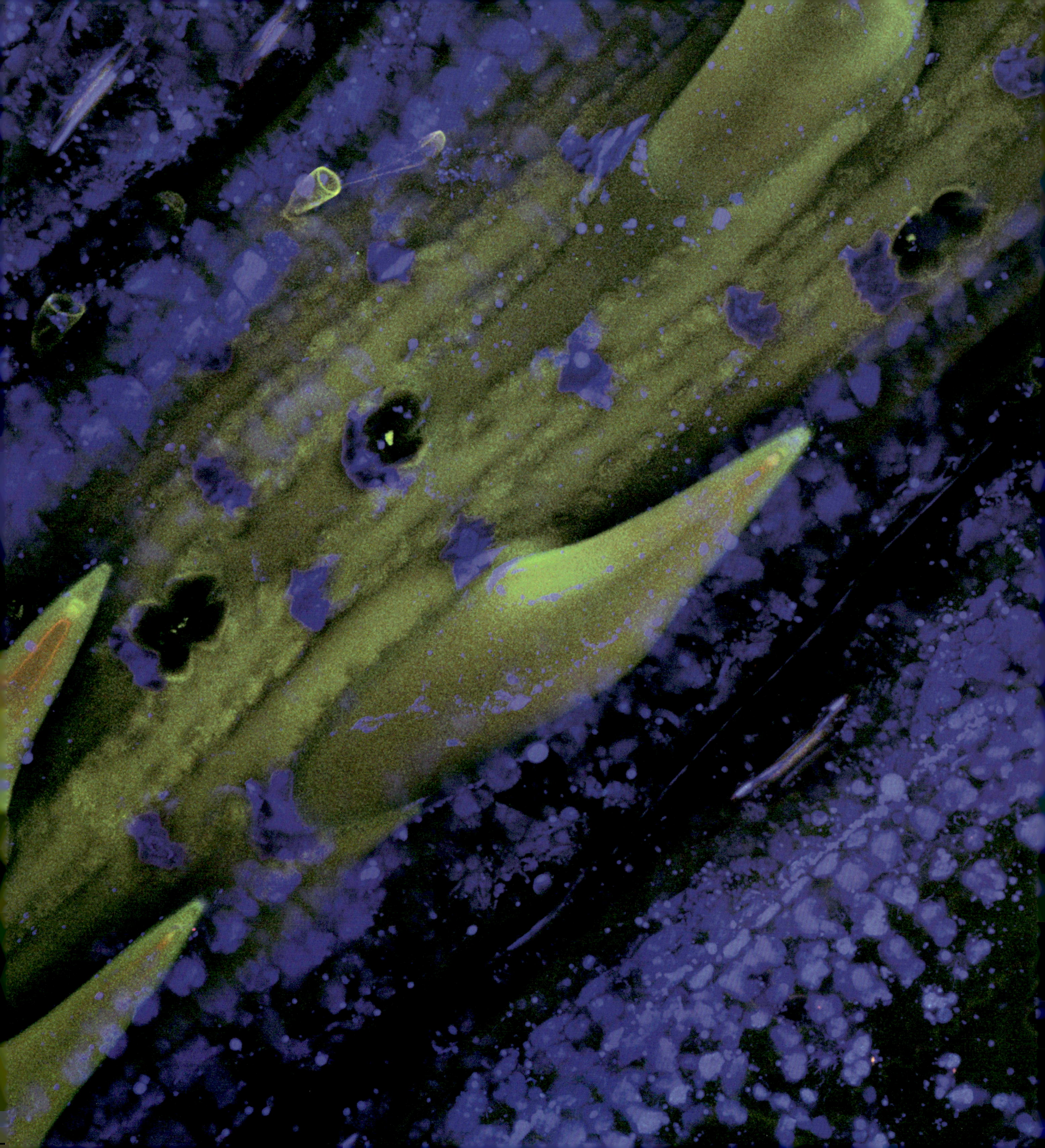

Poaceae

Zea luxurians

Guatemalan teosinte, wild maize

An annual flowering plant, true grass and a teosinte, this is the ancestor of maize in Mesoamerica and grows in the wet tropical biome. The domestication process lasted thousands of years, but produced what Guatemalan novelist Miguel Ángel Asturias called "Men of Maize", ancient civilisations that are based on the ritual cultivation and consumption of this essential sacred plant in all its radiant biodiversity.

The website Native Seeds Search provides the following information about teosinte, whose name derives from the Nahuatl word for sacred corn (*teotl* + *cintli*): "teosinte is an extremely important crop, as it believed that the subspecies *parviglumis* is the wild progenitor of corn. About 9,000 years ago, teosinte grew wild, as a grass-like plant, with a grain in a tough shell that was dispersed only when ripe. About 9,000-6,000 years ago, ancient people began to develop *parviglumis* teosinte into a crop that more closely resembles what we know as corn. Its kernels started to grow without the tough shell, and humans domesticated this plant for its grain, changing the size and textures of the kernels. This mutation causing the loss of the shell meant that the plant could no longer grow wild in its current form, since the kernels were unprotected from predators such as birds. Through these interactions with humans, it is thought that corn developed into the plant it is now".

Schaefer and Furst, in their important study of Huichol (Wixárika) culture, have written eloquently about the sacred qualities of corn and its incarnations as revered entities, pointing out spiritual links between Amerindian peoples throughout Mesoamerica and North America: "maize is not only the most sacred and important of the food plants, but has multiple divine personalities, appearing as the Mother of Maize, whose animal form is the dove, and as her five daughters, each a different colour. In some stories, Yoáwima,

Zea luxurians (wild maize); confocal image

50 μm

Blue Maize, is the most sacred of all, just as she is among the Pueblo Indians of the American Southwest. The young Maize Goddess is also known as Niwétsika. If the maize plant is female, the individual ear is male, and both are personified as divine beings, just as they were by the Mexica, or Aztecs, of Central Mexico".

Corn itself, as a plant, is so intimately related to Huichol (Wixárika) traditions and social structures, that it becomes a fundamental analogy for human existence in relation to the natural world. As Anthony A. Shelton puts it, "the life history of the Huichol is directly comparable to that of the maize. The ceremonies of birth, baptism, maturation, and death parallel one another. Even life itself is similar, establishing the maize family as a metaphor for the Huichol family".

Especially interesting are the transformative connections between ostensibly disparate elements of the Huichol (Wixárika) world and the Indigenous perceptions of it. Denis Lemaistre's description of these links is profoundly poetic: "peyote, deer and maize are united by a network of close correspondences. Myth and ritual present to us a circle of metamorphoses in which each figure is the creator of others at the same time as it is created by them, like vessels open to infinity".

There is so much to say about corn. Wade Davis writes about how healers in Mesoamerica pick up kernels of maize and scatter them over the surface of a table: "in their pattern lay the future, and with each successive throw came further insights that together formed the prognosis".

Teosinte, as well as the corn that evolved from it (both of which are considered sacred throughout the Americas), are very important for what they are, but also for what they *aren't*, namely, GMO corn. Biotechnology companies such as Bayer, BASF, Dow AgroScience, DuPont Pioneer, Monsanto, and Syngenta market GMO seed and related products, including herbicides.

The merger of Bayer and Monsanto – worth $66 billion – enabled Bayer to drop the Monsanto name due to the negative publicity surrounding this company that is one of the most hated businesses in the United States.

The genetically-engineered seed that is praised by some as hardier, more nutritious and more drought- and pest-resistant than non-GMO corn, nevertheless raises many serious questions.

Are there potential health concerns when scientists change the structure of corn in ways that would not occur through natural development, infusing it with animal DNA, herbicides, and pesticides? Does GMO corn, for example, cause cancerous tumors?

Will the global predominance of GMO corn make farmers from the developing world dependent on international seed companies with exclusive patents on these genetically-modified organisms? Will genetically-engineered genes introduced in wild plants ultimately cause a reduction in biodiversity?

Could GMO corn influence public health in terms of antibiotic-resistant bacteria? Could changes in the pollen of GMO corn affect the development of non-GMO corn through unintended cross-pollination? These and other potentially consequential issues certainly merit further research.

Finally, it is worth mentioning the pioneering work of Monica Gagliano. As a result of her revolutionary experiments on plant-language, she offers the following somber conclusion: "by revealing the vegetal voice, corn had come to ask that we recognise our attempts at silencing plants, because humans have something of a track record for silencing those whose voice they do not want to hear".

As the wild ancestor of modern maize, teosinte "has served as a study model of evolutionary processes and even more as a potential source of genomic variation to introgress maize varieties suitable for both food and feed". A group of experts in biotechnology from Mexico led by Mariana Zavala-López conducted research on teosinte's phenolic profile in 2017 and concluded that "teosinte's diverse genomic material could serve as a platform for the development of new breeding programs to restore the desired ancestral characteristics without sacrificing the current traits of modern maize, especially in terms of productivity". In a strange undoing and remaking of ethnobotanical history, the scientists affirm the following: "the generation of maize-teosinte hybrids that fulfill yield requirements and kernel quality is promising, especially if the new maize genotypes maintain the teosinte's high nutrient and phytochemical compositions".

A team of researchers from India headed by S. Sahoo conducted genetic and plant-breeding studies with teosinte and maize and published the results in *Tropical Plant Biology* in 2021. In their introduction, these scientists write that the evolution under domestication of maize resulted in the loss

Zea luxurians (wild maize); *the ancestral corn of Mesoamerica*

of alleles that could help the plant adapt more effectively now to abiotic stresses (such as heat, cold, and drought) as well as biotic stresses (including pathogens and herbivorous organisms). This reduction of genetic diversity makes maize more vulnerable to the impacts of climate change. On the other hand, maize's wild progenitor, teosinte, say the authors, has "more variation, more allelic options for addressing biotic and abiotic stresses". Artificial selection, therefore, produced maize, a miracle plant, to be sure, but one that has lost adaptive genes that still exist in teosinte, whose many varieties "are cross compatible with maize and therefore wild alleles introgression can be achieved easily using classical breeding approaches". In their breeding experiments, the scientists attempted to improve certain agronomic traits of maize, including flowering time, leaf angle, number of ears per plant, brace root systems, ear and kernel characteristics, weed tolerance, low and excess soil moisture stress, nitrogen fixation, as well as resistance to diseases and insects, all of which potentially contribute to the diversification of maize germplasm.

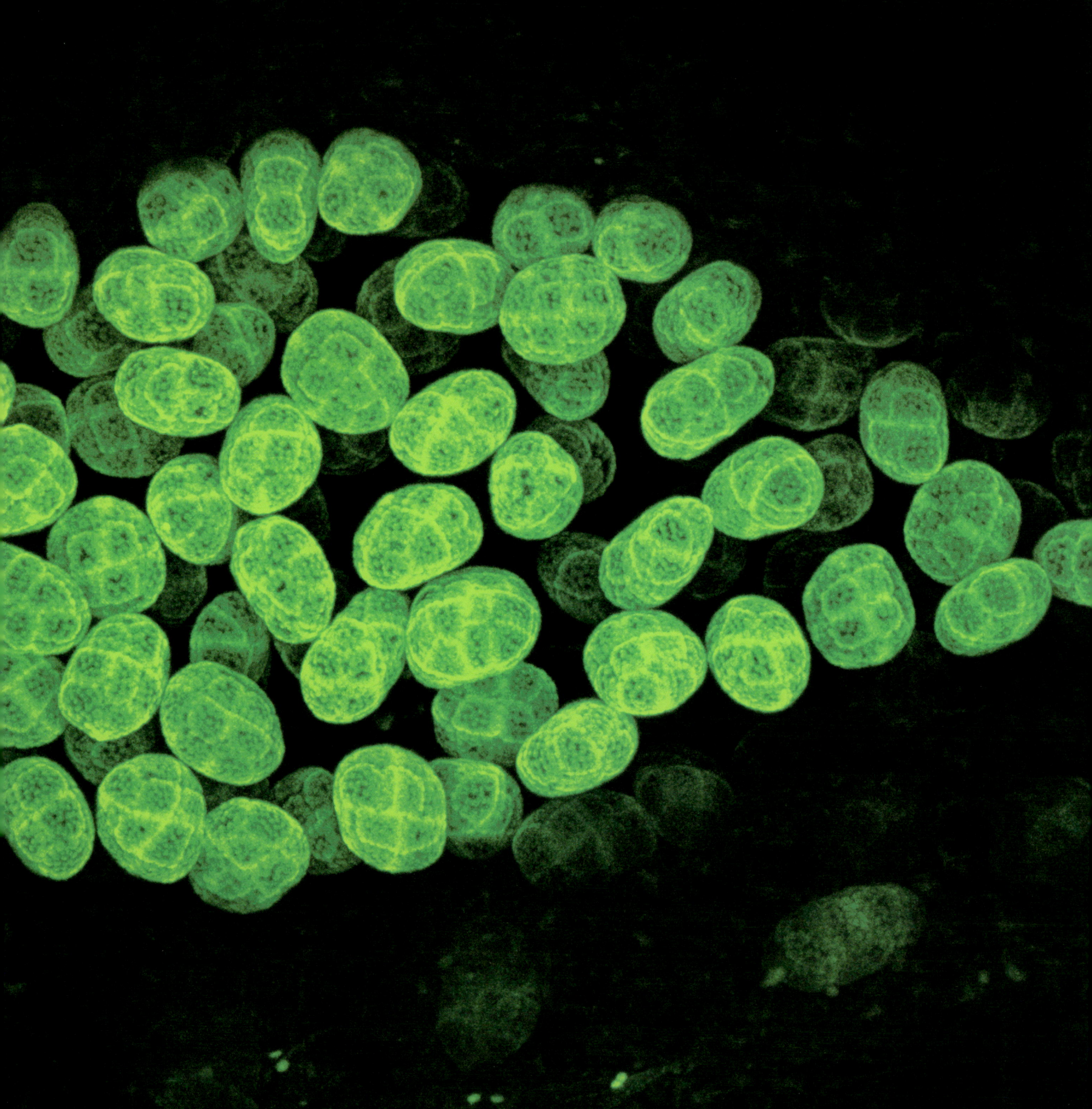

ABOUT THE AUTHORS

STEVEN F. WHITE

JILL PFLUGHEBER

Steven F. White was educated at Williams College and received his Ph.D. from the University of Oregon. White is the recipient of a Lannan Foundation residency and two Fulbright grants for a literary project in Chile and curricular development as a Senior Specialist in Nicaragua. When he was 22, his interest in sacred plants motivated him to visit a Cofan community in the Ecuadorean Amazon in 1977. During a transformative sabbatical year in 1993-94, he studied South American shamanism, and actively participated in the Santo Daime Church on the island of Santa Catarina in southern Brazil. He is the co-editor of *Ayahuasca Reader: Encounters with the Amazon's Sacred Vine* (Synergetic Press, 2016), which won an Independent Publishers Book Award. His essay on *Ceiba pentandra* appears in *The Mind of Plants: Narratives of Vegetal Intelligence*. He did an ecocritical study for his edition of *Seven Trees Against the Dying Light* by Nicaraguan poet Pablo Antonio Cuadra (Northwestern University Press, 2007), and translated the ethnobotanical poems of Esthela Calderón in *The Bones of My Grandfather* and *Paper Beehive* (Amargord, 2018 and 2022). He edited *El consumo de lo que somos: muestra de poesía ecológica hispánica contemporánea* (Amargord, 2014), and served as guest editor of a special issue on ecology and Latin American literature of *Review: Latin American Literature and the Arts* (2012). His research with Microscopy Specialist Jill Pflugheber *Microcosms: A Homage to Sacred Plants of the Americas* was presented as an exhibition at the Brush Art Gallery in 2020 at St. Lawrence University, where White taught Latin American literature and film for 34 years, and was a founder of the Caribbean and Latin American Studies interdisciplinary program. He currently serves on the editorial board of the journal *Plant Perspectives*.

Jill is a 1986 graduate of St. Lawrence University, with a strong background in biomedical research. She spent 17 years contributing to studies at Harvard University, University of Kentucky, and University of Texas Southwestern Medical Center. Her work has led to several journal publications.

In 2004, Jill returned to her alma mater, taking on the role of Microscopy Specialist. There, she taught courses in electron microscopy, confocal microscopy, and research methods in cell biology. One of the highlights of her teaching was the "Image of the Semester" contest in her confocal microscopy course. Students chose a favorite image from their portfolios, and anyone in the university community was invited to vote for their favorite. The winning images were then printed and displayed in the Launders Science Library for all to enjoy.

Jill's collaboration with Steven began after he saw and admired the contest images. Intrigued by the idea of what the leaf of *Banisteriopsis caapi* would look like under the confocal microscope, Steven reached out to Jill. One image turned into many, and, after more than four years of sample collection and imaging, the *Microcosms* collection was born. An exhibition of this initial work opened at St. Lawrence University's Richard F. Brush Art Gallery March 2, 2020, but was unfortunately cut short due to COVID-19 restrictions.

In mid-2023, Jill returned to biomedical research at the University of Kentucky, where she now manages Dr. Amelia Pinto's lab. She is actively involved in research exploring the impact of viruses and vaccines on the immune system.

Outside of work, Jill enjoys spending time with her family and horses. She frequently volunteers at events at Kentucky Horse Park and Masterson Station Park.

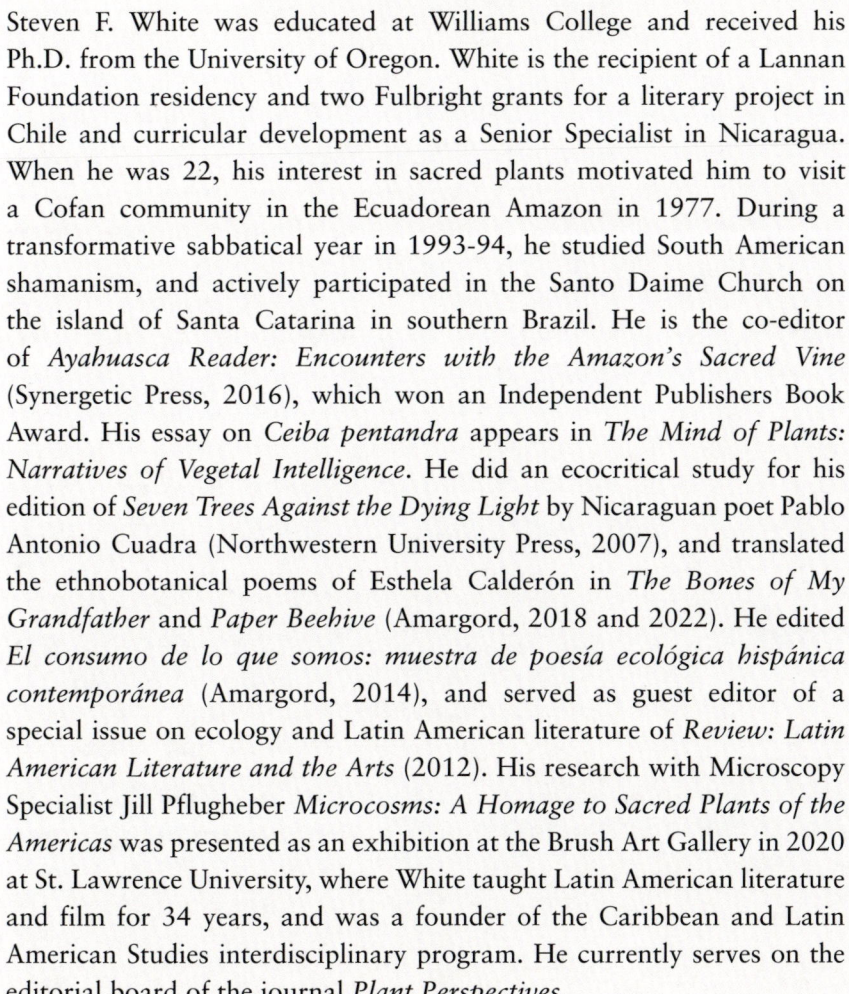

Anadenanthera colubrina (cebil); confocal image

50 μm

INDEX OF PLANTS & FUNGI ILLUSTRATED

Page	Latin name	Authority for name	Plant family
34, 36, 37	*Alicia anisopetala*	(A.Juss.) W.R.Anderson	Malpighiaceae
38, 40, 41	*Amaranthus cruentus*	L.	Amaranthaceae
42, 43, 46, 47, 48-49, 254	*Anadenanthera colubrina*	(Vell.) Brenan	Fabaceae
54	*Anadenanthera peregrina*	(L.) Speg.	Fabaceae
50, 52, 53	*Artemisia ludoviciana* subsp. *mexicana*	(Willd. ex Spreng.) D.D.Keck	Asteraceae
20, 24, 26, 33, 54, 58, 61, 74, 76, 77	*Banisteriopsis caapi*	(Spruce ex Griseb.) C.V.Morton	Malpighiaceae
frontis, 66, 75	*Banisteriopsis muricata*	(Cav.) Cuatrec.	Malpighiaceae
78, 81	*Bourreria huanita*	(Lex.) Hemsl.	Boraginaceae
back cover	*Brugmansia aurea*	Lagerh.	Solanaceae
23, 89, 91	*Brugmansia insignis*	(Barb.Rodr.) Lockwood ex R.E.Schult.	Solanaceae
82 88	*Brugmansia* × *candida* var. 'Culebra'	Pers.	Solanaceae
86, 87, 88, 90 92, 93	*Brugmansia sanguinea*	(Ruiz & Pav.) D.Don	Solanaceae
94, 97, 98, 99	*Brunfelsia grandiflora*	D.Don	Solanaceac
100, 102	*Bursera fagaroides*	(Kunth) Engl.	Burseraceae
104, 106, 107	*Calea ternifolia*	Kunth	Asteraceae
110, 111, 112, 113	*Cannabis sativa*	L.	Cannabaceae
front cover, 114, 116, 117, 118, 119	*Ceiba pentandra*	(L.) Gaertn.	Malvaceae
120, 122	*Cestrum parqui*	(Lam.) L'Hér.	Solanaceae
25, 27, 124, 126, 127	*Datura innoxia*	Mill.	Solanaceae
128, 131, 132, 133	*Desfontainia spinosa*	Ruiz & Pav.	Collumelliaceae
134, 136, 137	*Dianthera pectoralis*	(Jacq.) J.F.Gmel.	Acanthaceae
59, 64, 68, 70	*Diplopterys cabrerana*	(Cuatrec.) B.Gates	Malpighiaceae
67, 71	*Diplopterys longialata*	(Nied.) W.R.Anderson & C.Davis	Malpighiaceae
138, 141	*Drimys andina*	(Reiche) R.A.Rodr. & Quezada	Winteraceae
141	*Drimys winteri*	J.R.Forst. & G.Forst.	Winteraceae
142, 144, 145	*Erythroxylum novogranatense*	(D.Morris) Hieron.	Erythroxylaceae
146, 148, 149	*Heimia salicifolia*	Link	Lythraceae
150, 153	*Hierochloe odorata*	(L.) P.Beauv.	Poaceae
154, 158	*Ipomoea corymbosa*	(L.) Roth	Convolvulaceae
157, 158	*Ipomoea tricolor*	Cav.	Convolvulaceae
160, 161	*Latua pubiflora*	(Griseb.) Baill.	Solanaceae
162, 164, 165	*Leonotis nepetifolia*	(L.) R.Br.	Lamiaceae
29, 166, 168, 169, 170	*Lophophora williamsii*	(Lem. ex J.F.Cels) J.M.Coult.	Cactaceae
172, 174, 175	*Mimosa pudica*	L.	Fabaceae
176, 178, 179	*Mimosa tenuiflora*	(Willd.) Poir.	Fabaceae
180, 182, 186, 187	*Neltuma limensis*	(Benth.) C.E.Hughes & G.P.Lewis	Fabaceae
182	*Neltuma pallida*	(Humb. & Bonpl. ex Willd.) C.E.Hughes & G.P.Lewis	Fabaceae
188, 191	*Nicotiana rustica*	L.	Solanaceae
192, 194, 195	*Paullinia cupana*	Kunth	Sapindaceae
200	*Polylepis australis*	Bitter	Rosaceae
196, 199, 201	*Polylepis incarum*	(Bitter) M.Kessler & Schmidt-Leb.	Rosaceae
201	*Polylepis lanata*	(Kuntze) M.Kessler & Schmidt-Leb.	Rosaceae
201	*Polylepis reticulata*	Hieron	Rosaceae
200	*Polylepis tarapacana*	Phil.	Rosaceae
202, 205	*Psilocybe cubensis*	(Earl) Singer	Hymenogastraceae
62, 72	*Psychotria carthagenensis*	Jacq.	Rubiaceae
57, 62, 63, 73	*Psychotria viridis*	Ruiz & Pav.	Rubiaceae
206	*Salvia apiana*	Jeps.	Lamiaceae
208, 210, 211, 212	*Salvia divinorum*	Epling & Játiva	Lamiaceae
214, 218, 219	*Solandra maxima*	(Moc. & Sessé ex Dunal) P.S.Green	Solanaceae
222	*Tabernaemontana sananho*	Ruiz & Pav.	Apocynaceae
220, 223	*Tabernaemontana undulata*	Vahl	Apocynaceae
224, 225	*Tagetes lucida*	Cav.	Asteraceae
228, 231, 232, 233	*Theobroma cacao*	L.	Malvaceae
30, 234, 236-237	*Trichocereus macrogonus* var. *pachanoi*	(Britton & Rose) Albesiano & R.Kiesling	Cactaceae
238, 241, 242, 243	*Ullucus tuberosus* (*aborigineus*)	(Brücher) Sperling	Basellaceae
246, 247	*Virola calophylla*	(Spruce) Warb.	Myristicaceae
244, 248	*Virola theiodora*	(Spruce ex Benth.) Warb.	Myristicaceae
250, 253	*Zea luxurians*	(Durieu & Asch.) R.M.Bird	Poaceae

Common names	Native distribution in the Americas
Black ayahuasca, thunder ayahuasca	Argentina, Bolivia, Brazil, Paraguay, Peru
Alegría, amaranth, bledo, huautli	El Salvador, Guatemala, Honduras, Mexico, Nicaragua
Angico, cebil, vilca	Bolivia, Brazil, Argentina
Cohoba, parica, yopo	Antilles, Bolivia, Brazil, Colombia, Guianas, Venezuela
Estafiate, Mexican white sagebrush, western mugwort	Belize, El Salvador, Guatemala, Mexico, USA (AZ, CO, NM, NV, OK, TX)
Ayahuasca, hoasca, jagube, yagé, miiyabu	Argentina, Bolivia, Brazil, Colombia, Ecuador, Peru, Venezuela
Red ayahuasca	Argentina, Bolivia, Brazil, C. America (excl. Belize), Colombia, Ecuador, Guianas, Mexico, Para., Peru, Venez.
Esquisúchil, guie xoba, ik'al te, jazmín de Oaxaca, popcorn flower	Belize, Costa Rica, El Salvador, Guatemala, Honduras, Mexico, Nicaragua
Angel's trumpet, floripondio, huanduj, pehí, toé	Colombia, Ecuador, Venezuela
Angel's trumpet, floripondio, huanduj, pehí, toé	Bolivia, Brazil, Colombia, Ecuador, Peru
Culebra borrachero, lengua de tigre, mutscuai borrachero	Colombia, Ecuador
Blood-red angel's trumpet, misha toro, puka wantuk	Bolivia, Chile, Colombia, Ecuador, Peru
Chiric-sanango, chiri kaspi, ujajái, yesterday-today-tomorrow	Bolivia, Brazil North, Colombia, Ecuador, Peru, Venezuela
Copal, Mexican frankincense, pom, torchwood copal	Mexico, USA (AZ)
Aztec dream grass, thlé-pelakano, zacatechichi	Central America (excl. Panama), Colombia, Mexico
Marijuana, monte, pot, weed	[not native to any Americas]
Ceiba, kapok, lupuna, silk cotton tree, yax che-el cab	Bahamas, Bolivia, Brazil, C. America, Colombia, Ecuador, Antilles, Guianas, Mex., Peru, Carib., Venezuela
Chilean cestrum, Chilean flowering jessamine, hediondilla, palqui	Argentina, Bolivia, Brazil, Chile, Paraguay, Uruguay
Chamico, toloache	Mexico, USA (TX)
Borrachero de páramo, chapico, michai blanco, taique, trau-trau	Peru
Carpenter bush, chambá, chapantye, mashi hiri	Argentina, Boliv., Brazil, C. America, Colombia, Ecuador, Guianas, Mex., Parag., Peru, USA (FL), Venezuela
Chagropanga, chaliponga, oco-yagé	Brazil, Colombia, Ecuador, Peru, Venezuela
Huambisa	Bolivia, Colombia, Ecuador, Peru
Canelo enano, canelo andino, dwarf winter's bark	Argentina, Chile
Canelo, foye, winter's bark	Argentina, Chile
Coca	Colombia, Ecuador, Peru, Trinidad-Tobago, Venezuela
Sinicuichi, sun-opener	Argentina, Bolivia, Brazil, Jamaica, Mexico, Paraguay, Uruguay, USA (TX)
Holy grass, ohonte wenserákon, sipátsimo, sweetgrass, wingaashk	Canada, USA (excl. AL, AR, FL, GA, HI, KS, KY, LA, MS, MO, OK, SC, TN, TX, VA)
Christmas vine, ololiuhqui, snake plant, x-táabentun	Bahamas, Bolivia, Brazil, C. America, Colombia, Ecuador, Antilles, Mexico, Paraguay, Peru, Venezuela
Badoh negro, heavenly blue morning glory, tlililtzin	Mexico
Arbol de los brujos, kalku-mamüll, latué, latuy	Chile
Christmas candlestick, klipp dagga, lion's ear	[not native to any Americas]
Hikuli, peyote	Mexico, USA (TX)
Dormilona, morí viví, punyo-sisa, sensitive plant, touch-me-not	Bolivia, Brazil, Carib., C. America (excl. Panama), Colombia, Ecuador, Guianas, Mexico, Peru, Venezuela
Jurema, jurema preta, tepezcohuite	Brazil, Colombia, El Salvador, Honduras, Mexico, Nicaragua, Panamá, Venezuela
Algarrobo, espino negro, huarango, mesquite	Peru
Algarrobo, espino negro, huarango, mesquite	Bolivia, Colombia, Ecuador, Galápagos, Peru
Antzil moy, mapacho, piciyetl, sairi, tabaco, tobacco, ye, yetl	Peru
Guaraná	Brazil, Colombia, Ecuador, Guyana, Peru, Venezuela
Lampaya, queñua	Argentina
Lampaya, queñua	Bolivia
Lampaya, queñua	Bolivia
Lampaya, queñua	Ecuador
Lampaya, queñua	Bolivia, Chile
Di-shi-tjo-le-rra-ja, san isidro, magic mushroom, niños santos	Antilles, Central America, South America (excl. Chile, Guianas, Uruguay), USA
Amyruca	Antilles, Argentina, Bolivia, Brazil, Colombia, Ecuador, Guianas, Mexico, Paraguay, Peru, Uruguay, Venez.
Chacruna, kawa	Antilles, Bolivia, Brazil, Central America (excl. Honduras), Colombia, Mexico, Peru, Venezuela
Bee sage, buffalo sage, California sage, pellytaay, white sage	Mexico, USA (CA)
Diviner's sage, hierba de la pastora, ska pastora	Mexico
Chalice vine, kiéri, hueipatl	Central America, Colombia, Mexico
Kunakip, sikta, uchu sanango	Brazil, Colombia, Ecuador, Guianas, Peru, Venezuela
Becchete, mana heins	Brazil, Colombia, Costa Rica, Ecuador, Guianas, Panamá, Peru, Trinidad-Tobago, Venezuela
Cempoal, flor de los muertos, Mexican marigold, pericón, yauhtli	Central America, Mexico
Cacahuatl, cacao	Brazil, Colombia, Costa Rica, Ecuador, Guianas, Peru, Suriname, Venezuela
Aguacolla, huachuma, lapitoj, San Pedro cactus	Ecuador, Peru
Ancestral potato, melloco, olluko, papa lisa	Argentina, Bolivia, Peru
Cumala, epená, yãkoana	Bolivia, Brazil, Colombia, Ecuador, Guyana, Peru, Suriname, Venezuela
Cumala, epená, yãkoana	Brazil, Colombia, Venezuela
Guatemalan teosinte, wild maize	El Salvador, Guatemala, Honduras, Mexico

BIBLIOGRAPHY

Bibliography (abridged)
For the complete list of sources mentioned in
Microcosms – Sacred Plants of the Americas, **please use the QR Code:**

Aloi, Giovanni, *Why Look at Plants?: The Botanical Emergence in Contemporary Art*, Leiden and Boston (Brill, 2018)
"Botanical Decolonization in Defense of Cultivars," in Giovanni Aloi and Michael Marder, eds. *Vegetal Entwinements in Philosophy and Art*. Cambridge, Massachusetts: The MIT Press, 2023. pp. 276-289.
Arthur, Paul Longley and John Charles Ryan. "Tracing the Digital Plant Humanities: Narratives of Botanical Life and Human Flora Relations." *Future Humanities* 2 (3) (June 2024): https://doi.org/10.1002/fhu2.15
Bacigalupo, Ana Mariella. *Shamans of the Foye Tree: Gender, Power and Healing Among Chilean Mapuche*. Austin: University of Texas Press, 2007.
Barbira-Freedman, Françoise. "Shamanic Plants and Gender in the Healing Forest," in Elisabeth Hsu and Stephen Harris, eds. *Plants, Health and Healing: On the Interface of Ethnobotany and Medical Anthropology (Epistemologies of Healing, 6)*. New York and Oxford: Berghahn Books, 2010: 135-178.
"Tobacco and Shamanic Agency in the Upper Amazon: Historical and Contemporary Perspectives," in Andrew Russell and Elizabeth Rahman, eds. *The Master Plant: Tobacco in Lowland South America*. London and New York: Bloomsbury Academic, 2015: pp. 63-86.
Beilin, Katarzyna Olga and Sainath Suryanarayanan. "The War between Amaranth and Soy: Interspecies Resistance to Transgenic Soy Agriculture in Argentina." *Environmental Humanities* 9.2 (2017): 204-229.
Beilin, Kata. "The World According to Amaranth: Interspecies Memory in Tehuacán Valley," Environmental Cultural Studies Through Time: The Luso-Hispanic World *Hispanic Issues On Line* 24 (2019): 144-167.
Benítez, Guillermo, Martí March-Salas, Alberto Villa-Kamel, Ulises Cháves-Jiménez, Javier Hernández, Nuria Montes-Osuna, Joaquín Moreno-Chocano, Paloma Cariñanos, "The Genus *Datura* L. (Solanaceae) in Mexico and Spain – Ethnobotanical Perspective at the Interface of Medical and Illicit Uses," *Journal of Ethnopharmacology* 219 (2018): 133-151.
Benjamin, Walter. "News about Flowers," in *The Work of Art in the Age of Its Technological Reproducibility, and Other Writings on Media*. Cambridge and London: Belknap Press of Harvard University Press, 2008: 271-273.
Bennett, Bradley C. "Hallucinogenic Plants of the Shuar and Related Indigenous Groups in Amazonian Ecuador and Peru," *Brittonia* 44 (October 1992): 483-493.
Bennett, Bradley C. and Rocío Alarcón. "Hunting and Hallucinogens: The Use of Psychoactive Plants to Improve the Hunting Ability of Dogs." *Journal of Ethnopharmacology* 171 (2 August 2015): 171-183.
Beresford-Jones, David G. *The Lost Woodlands of Ancient Nasca: A Case-Study in Ecological and Cultural Collapse*. Oxford: University of Oxford Press, 2011.
Blakinger, John R. *Gyorgy Kepes: Undreaming the Bauhaus*. Cambridge and London: MIT Press, 2019.
Blossfeldt, Karl. *Masterworks*. Edited by Ann and Jürgen Wilde. Foreword and Botanical Notes by Hansjörg Küster. New York: D.A.P., 2017.
Boza Espinoza, Tatiana Erika and Michael Kessler. "A Monograph of the Genus *Polylepis* (Rosaceae)". *PhytoKeys* 203 (2022): 1–274.
Buhner, Stephen Harrod. *Sacred Plant Medicine: The Wisdom in Native American Herbalism*. Rochester, VT: Bear & Co., 2006.

Calderón, Esthela. *Soplo de corriente vital* (poemas etnobotánicos). Managua: 400 Elefantes, 2008.
"Corn," in *The Mind of Plants: Narratives of Vegetal Intelligence*. John C. Ryan, Patricia Vieira, and Monica Gagliano, eds. Santa Fe, New Mexico: Synergetic Press, 2021: 139-147.
Calvo, Paco with Natalie Lawrence. *Planta Sapiens: Unmasking Plant Intelligence*. London: The Bridge Street Press, 2022.
Casselman, Ivan et al. "From Local to Global: Fifty Years of Research on *Salvia divinorum*." *Journal of Ethnopharmacology* 151 (2014): 768-783.
Castaño-Uribe, Carlos and Thomas Van der Hammen, eds. *Arqueología de visiones y alucinaciones del cosmos felino y chamanístico de Chiribiquete*. Bogotá: Parques Nacionales Naturales de Colombia, 2006.
Castaño-Uribe, Carlos. "Simbología y cosmogonía en el arte rupestre de la Tradición Cultural Chiribiquete (TCC): una aproximación al Universo Chamanístico de los hombres jaguar," in Carlos Castaño-Uribe and Thomas Van der Hammen, eds. *Arqueología de visiones y alucinaciones del cosmos felino y chamanístico de Chiribiquete*. Bogotá: Parques Nacionales Naturales de Colombia, 2006: 83-163.
Chiribiquete: la maloka cósmica de los hombres jaguar. Bogotá: SURA, 2020.
Chacruna Institute for Psychedelic Plant Medicines (chacruna.net). (An amazingly complete website covering Spirituality, Culture, Science, Policy, Integration, News, and much more.)
Coccia, Emanuele. *The Life of Plants: A Metaphysics of Mixture*. Medford, MA: Polity Press, 2019.
Cuadra, Pablo Antonio. *Siete árboles contra el atardecer*. Caracas: Ediciones de la Presidencia de la República, 1980.
Seven Trees against the Dying Light. Translated by Greg Simon and Steven F. White. Evanston, Illinois: Northwestern University Press, 2007.
Davis, Wade. *One River: Explorations and Discoveries in the Amazon Rain Forest*. New York: Simon and Schuster, 1996.
"The Divine Leaf of Immortality," (on Coca) in *Beneath the Surface of Things: New and Selected Essays*. Vancouver: Greystone Books, 2024: 160-194.
Dixon, Roland B. "Words for Tobacco in American Indian Languages," *American Anthropologist New Series* 23.1 (January-March 1921): 19-49.
Ecuador Constitution (Rights of Nature). https://pdba.georgetown.edu/Constitutions/Ecuador/ecuador08.html (in Spanish)
https://pdba.georgetown.edu/Constitutions/Ecuador/english08.html (in English)
Estrada, Alvaro. *Vida de María Sabina: sabio de los hongos*. México, D.F.: Siglo XXI Editores, 1977.
María Sabina: Her Life and Chants. Translation and Commentaries by Henry Munn. Santa Barbara, CA: Ross-Erikson, 1981.
Favaron, Pedro. *Las visiones y los mundos: sendas visionarias de la Amazonía occidental*. Lima: CAAAP, 2017.
La senda del corazón: sabiduría de los pueblos indígenas de Norteamérica. México: DR Cooperativa de Producción y Servicios Editoriales, Heredad, S.C. de C.V., 2020.
"Desde el corazón de la cultura: filosofía, investigación académica y racionalidad poética de las naciones amerindias de la Amazonía," in Chacón, Gloria Elizabeth, Juan G. Sánchez Martínez and Lauren Beck, eds. *Abiayalan Pluriverses: Bridging Indigenous Studies & Hispanic Studies*. Amherst, MA: Amherst College Press, 2023: pp. 83-101.
Favaron Peyón, Pedro Martín. "*Netebaon joi*: la semiótica cósmica shipibo-konibo." *América sin Nombre* 32 (2025): 16-32.
Feldman Gracia, Leonardo. *El cacto San Pedro: su función y significado en Chavín de Huantar y la tradición religiosa de los Andes centrales*. Tesis: Magister en Arqueología Andina, Universidad Nacional Mayor de San Marcos, Lima, Perú (2006).
Flannery, Maura C. *In the Herbarium: The Hidden World of Collecting and Preserving Plants*. New Haven: Yale University Press, 2023.
Gagliano, Monica. *Thus Spoke the Plant*. Berkeley: North Atlantic Books, 2018.
Gagliano, Monica, John C. Ryan and Patricia Vieira, eds. *The Language of*

Plants: Science, Philosophy, Literature. Minneapolis: University of Minnesota Press, 2017.

Gates, Bronwen. "*Banisteriopsis, Diplopterys (Malpighiaceae)*". *Flora Neotropica* 30 (Feb. 18, 1982): 1-237. (Published by New York Botanical Garden Press on behalf of Organization for Flora Neotropica).

Gearin, Alex K. *Global Ayahuasca: Wondrous Visions and Modern Worlds*. Stanford: Stanford University Press, 2024.

Gebhart-Sayer, Angelika. "Design Therapy," in Luis Eduardo Luna and Steven F. White, eds. *Ayahuasca Reader: Encounters with the Amazon's Sacred Vine*. Santa Fe, NM: Synergetic Press, 2016: 217-223.

Gifford, Don. *The Farther Shore: A Natural History of Perception*. New York: The Atlantic Monthly Press, 1990.

Giraldo Herrera, César E. *Microbes and Other Shamanic Beings*. Cham, Switzerland: Palgrave Macmillan/Springer Nature, 2018.

Goethe, Johann Wolfgang von. *The Metamorphoses of Plants*. Introduction and Photography by Gordon L. Miller. Cambridge and London: MIT Press, 2009.

González Chévez, Lilián. "*Hueytlacatzintli*: enteógeno sagrado entre los nahuas de Guerrero," *Cuicuilco* 19 (53) (2012): 301-324.

González Romero, Osiris Sinuhé. "Decolonizing the Philosophy of Psychedelics", in *Philosophy and Psychedelics: Frameworks for Exceptional Experience*. Christine Hauskeller and Peter Sjöstedt-Hughes, (eds). London: Bloomsbury, 2022: 77-94.

"Sabiduría y ritual de los hongos sagrados en Mesoamérica", Revista Cultura y Droga, 28.35 (2023): 21-49.

"Cognitive Liberty and the Psychedelic Humanities," *Frontiers in Psychology* 14 (May 2023).

Haeckel, Ernst. *Art Forms in Nature*. Munich and New York: Prestel, 1998.

Hallé, Francis. *In Praise of Plants*. Portland, OR: Timber Press, 2002.

Hay, Alistair, Monika Gottschalk and Adolfo Holguín. *Huanduj: Brugmansia*. Surrey: Royal Botanic Gardens, Kew, 2012.

Kepes, György, ed. *The New Landscape in Art and Science*. Chicago: Paul Theobald, 1956.

Kesseler, Rob and Madeline Harley, eds. *Pollen: The Hidden Sexuality of Flowers*. Winterbourne, Berkshire, UK: Papadakis, 2014.

Kesseler, Rob and Wolfgang Stuppy. *Seeds: Time Capsules of Life*. Winterbourne, Berkshire, UK: Papadakis, 2024.

Kimmerer, Robin Wall. *Braiding Sweetgrass: Indigenous Wisdom, Scientific Knowledge and the Teaching of Plants*. Minneapolis: Milkweed Editions, 2013.

"Learning the Grammar of Animacy," *Anthropology of Consciousness* 28.2 (Fall 2017): 128-134.

The Serviceberry: Abundance and Reciprocity in the Natural World. New York: Scribner, 2024.

Kirkham, Nin and Chris Letheby. "Pyschedelics and Environmental Virtues," *Philosophical Psychology* (2022): DOI: 10.1080/09515089.2022.2057290

Kohn, Eduardo. *How Forests Think: Toward an Anthropology Beyond the Human*. Berkeley and Los Angeles: University of California Press, 2013.

Kopenawa, Davi and Bruce Albert. *The Falling Sky: Words of a Yanomami Shaman*. Translated by Nicholas Elliott and Alison Dundy. Cambridge, MA: Belknap Press, 2013.

Langdon, E. Jean. "Yagé among the Siona: Cultural Patterns in Visions," in David L. Browman, ed. *Spirits, Shamans, Stars: Perspectives from South America*. Berlin: De Gruyter Mouton, 1980: 63-80.

"Las clasificaciones del yagé dentro del grupo Siona: Etnobotánica, etnoquímica e história." *América Indígena* 46 (1986): 101–116.

"El valor de la narrativa: memoria y patrimonio entre los Siona," *Revista Del Museo de Antropología* 11 (2018): 91–100.

Langdon, Esther Jean, Laffay, Tom & Maniguaje-Yaiguaje, Pablo. "Resistencias, re-existencias y prácticas chamánicas: las poéticas y políticas de una visión contemporánea." *Mundo Amazónico* 14.1 (2023): 33-48.

The Last Forest, Directed by Luiz Bolognesi, performance by Davi Kopenawa, Netflix app, 2021.

Lema, Verónica S. "Contemporary Uses of Vilca (*Anadenanthera colubrina* var *cebil*): A Major Ritual Plant in the Andes" *Plants* 13.17: 2398 (2024).

Lockwood, T. E. "Generic Recognition of *Brugmansia*," *Botanical Museum Leaflets, Harvard University* 23 (1973): 273-284.

"The Ethnobotany of *Brugmansia*," *Journal of Ethnopharmacology* 1 (1979): 147-164.

Luna, Luis Eduardo. "*Ayahuasca*: A Powerful Epistemological Wildcard in a Complex, Fascinating and Dangerous World," in Dennis J. McKenna, ed. *Ethnopharmacologic Search for Psychoactive Drugs*, vol II, Santa Fe, New Mexico: Synergetic Press, 2017: pp. 24-35.

Luna, Luis Eduardo and Pablo Amaringo. *Ayahuasca Visions: The Religious Iconography of a Peruvian Shaman*. Berkeley: North Atlantic, 1991.

Luna, Luis Eduardo and Steven F. White, eds. *Ayahuasca Reader: Encounters with the Amazon's Sacred Vine*. Santa Fe, NM: Synergetic Press, 2000. 2nd revised and expanded edition, 2016.

Mancuso Stefano. *The Revolutionary Genius of Plants: A New Understanding of Plant Intelligence and Behavior*. New York: Atria Books, 2017.

Mangan, Arty. "The Sacred Plant Biocultural Recovery Initiative: An Interview with Gary Paul Nabhan." *Bioneers* (June 13, 2024): https://bioneers.org/the-sacred-plant-biocultural-recovery-initiative-an-interview-with-gary-paul-nabhan/?mc_cid=7fc4740671&mc_eid=e1bf4792a8

Maqueda, Ana Elda. "The Use of *Salvia divinorum* from a Mazatec Perspective," in *Plant Medicines, Healing and Psychedelic Science* in Beatriz Caiuby Labate and Clancy Cavnar, eds, Cham, Switzerland: Springer, 2018: 55-70.

Marder, Michael. *Plant-Thinking: A Philosophy of Vegetal Life*. New York: Columbia University Press, 2013.

The Philosopher's Plant: An Intellectual Herbarium. New York: Columbia University Press, 2014.

Marder, Michael with Artworks by Anaïs Tondeur. *The Chernobyl Herbarium: Fragments of an Exploded Consciousness*. London: Open Humanities Press, 2016.

Marley, Christopher. *Biophilia*. New York: Abrams, 2015.

Mathiowetz, Michael D. and Andrew D. Turner, eds. *Flower Worlds: Religion, Aesthetics, and Ideology in Mesoamerica and the American Southwest*. Tucson: The University of Arizona Press, 2021.

McKenna, Dennis J. "An Unusual Experience with 'Hoasca': A Lesson from the Teacher," in Luis Eduardo Luna and Steven F. White, eds. *Ayahuasca Reader: Encounters with the Amazon's Sacred Vine*. Santa Fe, NM: Synergetic Press, 2016: 320-326.

McNeil, Cameron L. "The Flowery Mountains of Copan: Pollen Remains from Maya Temples and Tombs," in Michael D. Mathiowetz and Andrew D. Turner, eds. *Flower Worlds: Religion, Aesthetics, and Ideology in Mesoamerica and the American Southwest*. Tucson: The University of Arizona Press, 2021: 129-148.

Millard, Dale. "Broad Spectrum Roles of Harmine in *Ayahuasca*," in Dennis J. McKenna, ed. *Ethnopharmacologic Search for Psychoactive Drugs* vol II, Santa Fe, New Mexico: Synergetic Press, 2017: pp. 82-94.

Moholy-Nagy, Lázló. *The New Vision: From Material to Architecture*. New York: Brewer, Warren and Putnam, 1932.

Monteles, Ricardo. "*Eu venho da floresta*": A sustentabilidade das plantas sagradas amazônicas do Santo Daime. Universidade Federal do Amazonas (Manaus) (fevereiro 2020). Tese (Doutorado em Ciências do Ambiente e Sustentabilidade na Amazônia). https://tede.ufam.edu.br/bitstream/tede/7682/2/Tese_RicardoMonteles_PPGCASA.pdf

Narby, Jeremy. *The Cosmic Serpent: DNA and the Origins of Knowledge*. New York: Tarcher/Putnam, 1999.

Plant Teachers: Ayahuasca, Tobacco and the Pursuit of Knowledge. Novato, California: New World Library, 2021.

Ogunbodede, Olabode, Douglas McCombs, Keeper Trout, Paul Daley & Martin Terry. "New Mescaline Concentrations from 14 Taxa/Cultivars of Echinopsis spp. (Cactaceae) ('San Pedro') and Their Relevance to Shamanic Practice." *Journal of Ethnopharmacology* 131.2 (2010): 356–362.

Ott, Jonathan. *The* Cacahuatl *Eater, Ruminations of an Unabashed Chocolate Addict*. Vashon, WA: Jonathan Ott Books, 1985.

Pharmacotheon: Entheogenic Drugs, Their Plant Sources and History. Kennewick, WA: Natural Products, 1996.

Shamanic Snuffs, or Entheogenic Errhines. Solothurn, Schweiz: Entheobotanica, 2001.

Page, Joanna. *Decolonizing Science in Latin American Art*. London: UCL Press, 2021.

Payaguaje, Fernando. *El bebedor de yagé*. Shushufindi—Río Aguarico, Ecuador: Vicariato Apostólico de Aguarico, 1990. https://samorini.it/doc1/alt_aut/lr/payaguaje-el-bebedor-de-yaje.pdf

The Yagé Drinker. Quito: Cicame, 2007. https://maps.org/images/pdf/books/yagedrinker/portada-the_yage_drinker.pdf

Plowman, Timothy. "*Brunfelsia* in Ethnomedicine." *Botanical Museum Leaflets* 25.10 (December 1977): 289-320.

"The Identification of Coca (*Erythroxylem* species)," *Botanical Journal of the Linnean Society* 84.4 (June 1982): 329-353.

"The Ethnobotany of Coca (*Erythroxylem* spp., Erythroxylaceae)," *Advances in Economic Botany* 1 (1982): 62-111.

Plowman, Timothy, Lars Olof Gyllenhaal and Jan Erik Lindgren. "*Latua pubiflora*: Magic Plant from Southern Chile," *Botanical Museum Leaflets, Harvard University* 23.2 (November 12, 1971): 61-92.

Pollan, Michael. *How to Change Your Mind: What the New Science of Psychedelics Teaches Us about Consciousness, Dying, Addiction, Depression and Transcendence*. New York: Penguin, 2018.

This Is Your Mind on Plants. New York: Random House, 2021.

Press, Sara V. "Ayahuasca on Trial: Biocolonialism, Biopiracy, and the Commodification of the Sacred." *History of Pharmacy and Pharmaceuticals* 63.2 (January 2022): 328-353.

Rätsch, Christian. *The Encyclopedia of Psychoactive Plants: Ethnopharmacology and Its Applications*. Rochester, Vermont: Park Street press, 2005.

Rick, John W. and Verónica S. Lema, et al. "Pre-Hispanic Ritual Use of Psychoactive Plants at Chavín de Huantar, Peru," PNAS 122.19 (May 5, 2025): https://doi.org/10.1073/pnas.2425125122

Russell, Andrew and Elizabeth Rahman, eds. *The Master Plant: Tobacco in Lowland South America*. London and New York: Bloomsbury Academic, 2015.

Ryan, John C. *Posthuman Plants: Rethinking the Vegetal through Culture, Art, and Poetry*. Champaign, IL: Common Ground Research Networks, 2015.

Plants in Contemporary Poetry: Ecocriticism and the Botanical Imagination. London: Routledge, 2017.

Ryan, John C. "On Being Called by Plants: Phytopoetics and the Phytosphere," *Plant Perspectives* 1.2 (2024): 258-275.

Ryan, John C., Patricia Vieira, and Monica Gagliano, eds. *The Mind of Plants: Narratives of Vegetal Intelligence*. Santa Fe, New Mexico: Synergetic Press, 2021.

Salmón, Enrique. *Iwígara (The Kinship of Plants and People): American Indian Ethnobotanical Traditions and Science*. Portland: Timber Press, 2020.

Schaaf, Larry J. *Sun Gardens: Cyanotypes by Anna Atkins*. New York: New York Public Library, 2018.

Schaefer, Stacy B. and Peter T. Furst, eds. *People of the Peyote: Huichol Indian History, Religion & Survival*. Albuquerque: University of New Mexico Press, 1996.

Schlanger, Zoë. *The Light Eaters: How the Unseen World of Plant Intelligence Offers a New Understanding of Life on Earth*. New York: Harper, 2024.

Schultes, Richard Evans, Albert Hofmann and Christian Rätsch. *Plants of the Gods: Their Sacred, Healing and Hallucinogenic Powers*. Rochester, VT: Healing Arts Press, 2001.

Sheldrake, Merlin. *Entangled Life: How Fungi Make Our Worlds, Change Our Minds, and Shape our Futures*. London: The Bodley Head, 2020.

Stamets, Paul. *Psilocybin Mushrooms in Their Natural Habitats: A Guide to the History, Identification and Use of Psychoactive Fungi*. Berkeley: Ten Speed Press, 2025.

Stamets, Paul. *Psilocybin Mushrooms of the World: An Identification Guide*. Berkeley: Ten Speed Press, 1996.

Stewart, Susan. *On Longing: Narratives of the Miniature, the Gigantic, the Souvenir, the Collection*. Baltimore: Johns Hopkins University Press, 1984.

Strüwe, Carl. *Formen Des Mikrokosmos: Gestalt Und Gestaltung Einer*. Munich: Prestel, 1955.

Sugden, Laurel Anne and Josip Orlovac Del Río. "The Ark: Biocultural Sustainability for the San Pedro Cactus" in *Ethnopharmacologic Search for Psychoactive Drugs (Vol. 3): Proceedings from the 2022 Conference*. Forthcoming, Synergetic Press, 2025.

Tedlock, Dennis. *Popol Vuh: The Definitive Edition of the Mayan Book of the Dawn of Life and the Glories of Gods and Kings*. New York: Simon & Schuster/Touchstone, 1996.

Terranova, Charissa N. *Art as Organism: Biology and the Evolution of the Digital Image*. London and New York: I. B. Taurus, 2016.

Terranova, Charissa N. and Meredith Tromble, eds. *The Routledge Companion to Biology in Art and Architecture*. New York and London: Routledge, 2017.

Torres, Constantino Manuel and David B. Repke. *Anadenanthera: Visionary Plant of Ancient South America*. Binghamton: The Haworth Herbal Press, 2006.

Torres, Constantino Manuel. Artes visuales de Donna Torres. *La ayahuasca, el yagé y otras bebidas psicoactivas: sus orígenes y su historia*. Santiago, Chile: CIIR/Pehuén, 2024.

Vickers, William T. and Timothy Plowman. "Useful Plants of the Siona and Secoya Indians of Eastern Ecuador." *Fieldiana Botany* (N.S.) 15 (1984):1–63.

Virdi, Jasmine. "The Scientific Study of Ayahuasca Ethno-Varieties with Regina Célia de Oliveira," Chacruna Institute for Psychedelic Plant Medicines. (April 23, 2021): https://chacruna.net/botany_varieties_ayahuasca_banisteriopsiscaapi/

Wang, Rui and A. Dobritsa. "Exine and Aperture Patterns on the Pollen Surface: Their Formation and Roles in Plant Reproduction," *Annual Plant Reviews* 1 (2018): 1-40.

Weisberger, Jonathan Miller. *Rainforest Medicine: Preserving Indigenous Science and Biodiversity in the Upper Amazon*. Berkeley: North Atlantic, 2013.

White, Steven F. "Ceiba," in *The Mind of Plants: Narratives of Vegetal Intelligence*. John C. Ryan, Patricia Vieira, and Monica Gagliano, eds. Santa Fe, New Mexico: Synergetic Press, 2021: 99-106.

Wilbert, Johannes. *Tobacco and Shamanism in South America*. New Haven: Yale University Press, 1987.

Williams, Keith J. et al. "Indigenous Philosophies and the 'Psychedelic Renaissance'," *Anthropology of Consciousness* 33.2 (2022): 506-527.

SELECT PURVEYORS OF PLANTS, SEEDS, & SNUFFS

Cactus Kingdom
Garden Shaman
Maui Green Nutraceuticals
Misplant
Sacred Succulents
Shamanic Supply USA
Soul Vine
Spirit of the Jungle/Casa del Colibrí
Waking Herbs
World Seed Supply

MICROCOSMS: THE INNER WORLDS OF SACRED PLANTS

FOOTNOTES

[1] Joanna Page, *Decolonizing Science in Latin American Art*. (London: UCL Press, 2021): 3.

[2] Robin Wall Kimmerer. (SUNY ESF/Center for Native Peoples and the Environment). Webinar: "The Fortress, the River and the Garden: A New Metaphor for Symbiosis between Indigenous and Scientific Knowledges," *Indigenous Education Institute/A Sense of Place: Indigenous Perspectives on Earth, Water and Sky series* (August 20, 2020), 1:22:58, video available on You Tube.

[3] John C. Ryan, "On Being Called by Plants: Phytopoetics and the Phytosphere," *Plant Perspectives* 1.2 (2024): 268.

[4] Ryan, p. 269.

[5] See Arty Mangan. "The Sacred Plant Biocultural Recovery Initiative: An Interview with Gary Paul Nabhan." *Bioneers* (June 13, 2024).

[6] Keith J. Williams, et al. "Indigenous Philosophies and the 'Psychedelic Renaissance'," *Anthropology of Consciousness* 33.2 (2022): 510. https://doi.org/10.1111/anoc.12161

[7] Charissa Terranova and Meredith Tromble. "Introduction" in Charissa Terranova and Meredith Tromble, eds. *The Routledge Companion to Biology in Art and Architecture*. (New York and London: Routledge, 2017): 4.

[8] Maura C. Flannery. *In the Herbarium: The Hidden World of Collecting and Preserving Plants*. (New Haven: Yale University Press, 2023): 249.

[9] Enrique Salmón. *Iwígara, the Kinship of Plants and People: American Indian Ethnobotanical Traditions and Science*. (Portland, Oregon: Timber Press, 2020): 10.

[10] Emanuele Coccia. *The Life of Plants: A Metaphysics of Mixture*. Translated by Dylan J. Montanari. (Cambridge: Polity Press, 2019): 36.

[11] Coccia, p. 47.

[12] Jonathan Ott. *The Cacahuatl Eater, Ruminations of an Unabashed Chocolate Addict*. (Vashon, WA: Jonathan Ott Books, 1985): 74.

[13] J. P. Hodin. "The Painter's Handwriting," in Gyorgi Kepes, ed. *Vision + Value Series: Sign, Image, Symbol*. (New York: George Braziller, 1966): 151.

[14] Zoë Schlanger. *The Light Eaters: How the Unseen World of Plant Intelligence Offers a New Understanding of Life on Earth*. (New York: Harper, 2024): 109.

[15] Rui Wang and A. A. Dobritsa. "Exine and Aperture Patterns on the Pollen Surface: Their Formation and Roles in Plant Reproduction," *Annual Plant Reviews* 1 (2018): 6.

[16] Rob Kesseler. "Pixillated Pollen," in Rob Kesseler and Madeline Harley, eds. *Pollen: The Hidden Sexuality of Flowers*. (London, UK: Papadakis, 2004): 183.

[17] Kesseler, p. 185.

[18] Dennis Tedlock, editor and translator. *Popol Vuh: The Definitive Edition of the Mayan Book of the Dawn of Life and the Glories of Gods and Kings*. New York: Simon & Schuster, 1996: p. 21.

[19] Tedlock, p. 218.

[20] Michael Marder. *Plant-Thinking: A Philosophy of Vegetal Life*. New York: Columbia University Press, 2013: p. 2.

[21] Marder, p. 3.

WORKS CITED

Coccia, Emanuele. *The Life of Plants: A Metaphysics of Mixture*. Translated by Dylan J. Montanari. Cambridge: Polity Press, 2019.

Flannery, Maura C. *In the Herbarium: The Hidden World of Collecting and Preserving Plants*. New Haven: Yale University Press, 2023.

Hodin, J. P. "The Painter's Handwriting," in Gyorgi Kepes, ed. *Vision + Value Series: Sign, Image, Symbol*. (New York: George Braziller, 1966): 150-167.

Kesseler, Rob. "Pixillated Pollen," in Rob Kesseler and Madeline Harley, eds. *Pollen: The Hidden Sexuality of Flowers*. Winterbourne, Berkshire, UK: Papadakis, 2014: 179-185.

Kimmerer, Robin Wall. (SUNY ESF/Center for Native Peoples and the Environment). Webinar: "The Fortress, the River and the Garden: A New Metaphor for Symbiosis between Indigenous and Scientific Knowledges," Indigenous Education Institute/A Sense of Place: Indigenous Perspectives on Earth, Water and Sky series (August 20, 2020), https://www.youtube.com/watch?v=gkG5QyHRmhc

Mangan, Arty. "The Sacred Plant Biocultural Recovery Initiative: An Interview with Gary Paul Nabhan." *Bioneers* (June 13, 2024).

Marder, Michael. *Plant-Thinking: A Philosophy of Vegetal Life*. New York: Columbia University Press, 2013.

Ott, Jonathan. *The* Cacahuatl *Eater, Ruminations of an Unabashed Chocolate Addict*. Vashon, WA: Jonathan Ott Books, 1985.

Page, Joanna. *Decolonizing Science in Latin American Art*. London: UCL Press, 2021.

Ryan, John C. "On Being Called by Plants: Phytopoetics and the Phytosphere," *Plant Perspectives* 1.2 (2024): 258-275.

Salmón, Enrique. *Iwígara, the Kinship of Plants and People: American Indian Ethnobotanical Traditions and Science*. Portland, Oregon: Timber Press, 2020.

Schlanger, Zoë. *The Light Eaters: How the Unseen World of Plant Intelligence Offers a New Understanding of Life on Earth*. New York: Harper, 2024.

Tedlock, Dennis, editor and translator. *Popol Vuh: The Definitive Edition of the Mayan Book of the Dawn of Life and the Glories of Gods and Kings*. New York: Simon & Schuster, 1996.

Terranova, Charissa and Meredith Tromble, eds. *The Routledge Companion to Biology in Art and Architecture*. New York and London: Routledge, 2017.

Wang, Rui and A. A. Dobritsa. "Exine and Aperture Patterns on the Pollen Surface: Their Formation and Roles in Plant Reproduction," *Annual Plant Reviews* 1 (2018): 1-40.

Williams, Keith J. et al. "Indigenous Philosophies and the 'Psychedelic Renaissance'," *Anthropology of Consciousness* 33.2 (2022): 506-527.

ACKNOWLEDGEMENTS

First and foremost, I would like to give thanks to the plants of *Microcosms* for all their guidance and support, their wisdom and shared breath. I've tried to be their good ally in these pages, and I have no doubt that they've been mine for many years on a sometimes harrowing path that required a steadfast spirit, loyalty, and resilience. It has been a privilege to care for these plant-relatives, co-become with them in close familiarity, and, now, share what accomplished Microscopy Specialist Jill Pflugheber and I were fortunate to be able to create for *Microcosms – Sacred Plants of the Americas* from Papadakis Publisher, a book that, in my case, builds on a lifetime of academic and spiritual interests.

In terms of live plant acquisitions necessary for working with a confocal microscope, Peter Wroblewski participated in this project in an amazingly generous way, as did Luis Eduardo Luna, Brian Richards, and Marc Bates. Select purveyors of plants, seeds, and derivative extracts and snuffs of interest to us in our project include: Cactus Kingdom, Garden Shaman, Heirloom Seeds, Maui Green Nutraceuticals, Misplant, Sacred Succulents, Shamanic Supply USA, Soul Vine, Spirit of the Jungle/Casa del Colibrí, Strictly Medicinal Seeds, Waking Herbs, and World Seed Supply.

An extra special recognition of gratitude goes to our dear friend, the talented poet, pastel artist, and plant whisperer Becky Harblin, who worked with us on *Microcosms* from its beginnings. The confocal images and some of the photos of *Amaranthus cruentus*, *Datura innoxia*, *Erythroxylem novogranatense*, *Hierochloe odorata*, *Lophophora williamsii* and *Trichocereus macrogonus* var. *pachanoi* are from plants she tended at her beautiful home in Norwood, New York. She passed on 9th April 2025, and started her journey into the Flower Worlds she loved.

We constructed this publication in stages. We want to acknowledge that plants we were able to identify as being culturally-significant for a wide range of Amerindian groups were scanned at St. Lawrence University's Microscopy and Imaging Center. This same institution (where Jill and I were colleagues) also facilitated the first *Microcosms* art exhibition in March 2020. In this regard, we want to express our appreciation to Catherine Tedford, Director of the Brush Art Gallery at St. Lawrence University, her staff, as well as to Josephine Skiff, Assistant Director of the Newell Center for Arts Technology, for their collective skill and persistence that enabled us to overcome certain challenges and make the first public showing of what I have called "microcosmic phytoformalism" a memorable event.

We would like to give our most profound thanks to Eric Williams-Bergen for his volunteered expertise to design and build the *Microcosms* website (with help from Eden Williams-Bergen and Jean Williams-Bergen). Over the last three years, Eden has been in charge of all technical aspects of updating the website, which will be used to support the Papadakis *Microcosms* book through a QR code. The *Microcosms* website is available in its entirety not only in English but also in French, German, Portuguese, and Spanish. Roy Caldwell and Ullrich Umann were instrumental in helping to ensure the quality of this diverse linguistic presence of *Microcosms*. We hope that the book will become available in these and other languages as future co-editions.

After the *Microcosms* website was launched on Earth Day in 2022, there were numerous distinguished thinkers from a range of disciplines who offered to write commentaries on our work that we published online. They include: Giovanni Aloi, Paco Calvo, Carlos Castaño-Uribe, Azucena Castro, Osiris Sinuhé González Romero, Stephen Grey, Rob Kesseler, Sara Lewis, Jorge Marcone, Dennis McKenna, Sharday Mosurinjohn, Jeremy Narby, Mark Plotkin, John C. Ryan, and many others.

Words can barely express how grateful we are to Professor Stephen J. Haggarty from Harvard Medical School (and Scientific Director of Neurobiology, Massachusetts General Hospital, Center for Neuroscience of Psychedelics), who, in collaboration with Harvard's Mahindra Humanities Center, and with the help of Surya Reis and Sunil Wadhwa, will be curating and sponsoring the exhibition "Mapping Cross-Cultural Connectivity through the Art & Science of Psychedelic Plants" at the Paul S. Russell Museum of Medical History and Innovation in Boston. The exhibition will feature confocal images from *Microcosms,* and the Papadakis book is to be launched on this occasion.

We were honoured to receive a great deal of enthusiastic encouragement and support that enabled us to spread the word about our work with plants. Patricia Lea Watts and Olivia Ann Carye Hallstein talked with us for *EcoArtSpace*. Paul Moss and Keith J. Williams did a program on *Microcosms* for the series sponsored by *The Plant Initiative*. It was a pleasure to have a long conversation with Kate Brelje on the podcast *Networking with Plants in the Anthropocene*. Joela Jacobs was always helpful in terms of creating connections between us and other members of the organization she co-founded *Literary and Cultural Plant Studies Network*. We really enjoyed seeing our work in German in *Lucys Rausch*, thanks to Markus Berger. We'd like to thank Donna Torres and Constantino Manuel Torres for their friendship and shared plant wisdom over the years. Manolo curated the exhibition *Shamanism: Visions Outside of Time* at the Museo Chileno de Arte Precolombino, and used work from *Microcosms* to publicize the event. It was incredible to see our Huachuma image become a giant street banner in downtown Santiago, Chile! Dr. Kata Beilin from the University of Wisconsin-Madison was kind enough to publish our article on *Microcosms* in the inaugural issue of LACIS on Inter/Transdisciplinarity. We're grateful to Benjamin De Loenen and his staff for supporting *Microcosms* in the publications of the International Center for Ethnobotanical Education, Research and Service (ICEERS), an organization we should *all* thank, especially for their magnificent efforts with the Ayahuasca Defense Fund. We have greatly enjoyed our collaborations with Dr. Liam Engel from *Entheogenesis Australis* and the journal *Kahpi*, which republished a brilliant article by Jonathon Miller Weisberger on Siekopai ancestral plant knowledge that originally appeared on the *Microcosms* website. Jonathon was kind enough to provide some superb photos of *Diplopterys cabrerana* for the *Microcosms* book. This is the right moment to thank Alan Rockefeller for permission to reproduce his breathtakingly precise photographs that document *Diplopterys longialata* (huambisa) in bloom.

And, speaking of photographers, *Microcosms* has benefited greatly from the generosity of experts with a gift for revealing the beautiful and striking lives of plants and trees that grow in the most diverse ecosystems: deserts, Amazonian jungles, and high Andean peaks. Here are the names of some who deserve our thanks for allowing us to reproduce their photographs: Thibaud Aronson, Tom Baldwin, Yarelis Benavides, James Benefield, Richo Cesco, Riley Fortier, the Ecuadorean Brugmansia specialists Adolfo Holguín and Eduardo Sánchez, Camilo Jarquín Calderón, Lizzy Kaya Leshy, Timothy Paine (suggested to me by Sue Willson), Francisca Pezo Sáez, Glenn Shepard, Jr., and María del Rosario Vicente Aquino.

I am thankful and honoured that my friends, the Peruvian artist Anderson Debernardi and the Colombian artist Jeisson Castillo, are also part of the *Microcosms* book.

We are deeply grateful that Neil Logan and his collaborators in Hawai'i offered us the privileged opportunity to work with different varieties of flowering *Banisteriopsis caapi* (beloved by Siekopai healers in Ecuador) such as Tara Yagé, Tzinca, and Wa'i Yagé. Neil also gave us an intensive class on the potential worldwide nutritional and ecological benefits of cultivating *Neltuma*. From Neil, we gained a fuller understanding of *Neltuma*'s role in the botanical history of ancient coastal Peru and, much later, Hawai'i.

Ben Kamm, the owner of *Sacred Succulents*, contributed many plant photographs for this publication as well as some writing on *Polylepis* and *Ullucus tuberosus* based on his extensive knowledge and demanding research trips throughout the Americas. I am very appreciative of his generosity and expertise, as well as his help in acquiring some important plants that enabled our work on the *Microcosms* project to proceed in a more comprehensive manner.

Thanks to David Hornung for permission to photograph his yarn paintings by Wixárika artists Antonio López Pinedo and José Benítez Sánchez, and to Yvonne Da Silva Negrín of the Wixárika Resource Center in Berkeley, California for allowing us to reproduce a work by Guadalupe González Ríos.

We want to say an immense thank you to Gusmano Cesaretti and his wife Rosa for permission to use his photograph of María Sabina and also for telling me the beautiful story of how he met the Mazatec healer in Huautla de Jiménez, Oaxaca, Mexico.

Personally, I am grateful to Professor Federico Roda for introducing me recently to Bernardo Chindoy, grandson of Salvador Chindoy, the Kamëntsá healer who collaborated with Richard Evans Schultes in Colombia's Sibundoy Valley decades ago. One small gesture of reciprocity is considering concrete ways that *Microcosms* might contribute to the establishment of a garden that conserves the rare brugmansias (that Roda calls "flying plants") and the endangered ethnobotanical knowledge of their stewards who use them for healing.

My publisher Alexandra Papadakis is far and away the best editor and designer of books that I have ever worked with and known. I truly believe that Papadakis Publisher is the best home in the world for *Microcosms – Sacred Plants of the Americas*. What a joy she and her assistant Molly Dewar have made this journey. Deep gratitude!

The University of Kentucky's Light Microscopy Core Facility (RRID:SCR_026405) generously granted access for Jill to use the proprietary software necessary to prepare these confocal images for publication. We gratefully acknowledge the NSF Award (ID 1626166) received by St. Lawrence University for "MRI: Acquisition of Confocal Laser Scanning Microscope for Research and Training in the Biological Sciences" that allowed us to do the confocal imaging for the *Microcosms* project.

The plants put my wife, the Nicaraguan poet and visual artist Esthela Calderón, on this path with me, and that has made all the difference. I can't thank her, or the plants, enough!

Steven F. White & Jill Pflugheber

PICTURE CREDITS

Back cover: © IBRESTER / Adobe Stock
Endpapers: artwork by an anonymous Shipiba artist; permanent collection, Brush Art Gallery, St. Lawrence University
Page 5: collection David Hornung
Page 8: courtesy of the Department of Special Collections, Stanford University Libraries; used with permission of the Estate of György Kepes
Page 11 (top): © Esthela Calderón
Page 11 (bottom): © Jill Pflugheber
Page 12 (top): © Art Collection 3, Alamy Stock Photo
Page 12 (bottom): courtesy of the Natural History Museum, London
Page 13 (top left): Wikimedia Commons
Page 13 (top right): © Carlo Bollo, Alamy Stock Photo
Page 13 (below): © Science History Images, Alamy Stock Photo
Page 14: © Peter Barritt, Alamy Stock Photo
Page 15 (top): © Peter Barritt, Alamy Stock Photo
Page 15 (bottom left): © Bill Waterson, Alamy Stock Photo
Page 15 (centre): courtesy of Lily Klee
Page 15 (right): © Artokoloro, Penta Springs Limited, Alamy Stock Photo
Page 16 (top): © billwissedition Ltd. & Co. KG, Alamy Stock Photo
Page 16 (below): © Heritage Image Partnership Ltd., Alamy Stock Photo
Page 17: by permission of Prestel-Verlag, München; photos © Esthela Calderón
Page 18: Santos Motoapohua de la Torre, photo © Fernando Alarriba
Page 37: © Tom Baldwin
Page 40 (top): © Kabar/Shutterstock
Page 40 (bottom): © Esthela Calderón
Page 43: © Constantino Manuel Torres
Page 45: © Donna Torres
Page 46: © Lucas Faramiglio/Shutterstock
Page 52: © John Lambing/Alamy
Page 53: © mediasculp/Alamy
Page 59: © Jonathon Miller Weisberger
Page 61: (top left): © Jonathon Miller Weisberger
Page 61: (bottom left): © Jonathon Miller Weisberger
Page 61: (right): courtesy of Jeisson Castillo; photo © Esthela Calderón
Page 62 (top): © Camilo Jarquín Calderón
Page 62 (centre): © Steven F. White
Page 62 (bottom): © Esthela Calderón
Page 65: (left and centre): © Tom Baldwin
Page 65 (right): courtesy of the family of Fernando Payaguaje
Page 66: © Jonathon Miller Weisberger
Page 67 and 71: © Alan Rockefeller
Page 68: anonymous Shipibo artist; collection of Esthela Calderón; photo © E. C.
Page 80: © Amovitania/Shutterstock
Page 81 (left): © staff member of the San Miguel Cathedral, Tegucigalpa, Honduras
Page 81 (right): © María del Rosario Vicente Aquino, iNaturalist
Page 85 (top left and right): courtesy of family of Richard Evans Schultes
Page 85 (above left): still from video © Lacifraimpar
Page 86: courtesy of Anderson Debernardi; photo © Esthela Calderón
Page 87: © Eduardo Sánchez
Page 88 (top and bottom left): © Adolfo Holguín
Page 88 (bottom right): © Eduardo Sánchez
Page 89: © Christian Vinces/Shutterstock
Page 99 (top): © Merin Elsa Jacob/Shutterstock
Page 99 (bottom): anonymous Shipiba artist; photo © Esthela Calderón
Page 102 (top and bottom centre): © Ben Kamm
Page 102 (bottom left and right): © Fabiano Sodi
Page 106: © Daldale/Shutterstock
Page 110: © bt_photo/Shutterstock
Page 112: © Melica/Shutterstock
Page 113: © Kym MacKinnon/Shutterstock
Page 114: © Camilo Jarquín Calderón

Page 116: © Steven F. White
Page 117: © Lizzy Kaya Leshy
Page 122 (left): © Furiarossa/Shutterstock
Page 122 (right): © Esthela Calderón
Page 126 (top): © Esthela Calderón
Page 126 (bottom): © AlyarMSD/Shutterstock
Page 130: courtesy of the Schultes family
Page 131 (top): © Ben Kamm
Page 131 (bottom): © Antoniya Kadiyska/Shutterstock
Page 136: © Esthela Calderón
Page 140: © Edmundo Stockins; courtesy of Wellcome Collection
Page 141 (top left and right): © Francisca Pezo Sáez
Page 141 (bottom): photo © Alex Manders/Shutterstock
Page 144: © Ben Kamm
Page 145 (top): © Greentellect Studio/Shutterstock
Page 145: © Ben Kamm
Page 148: © Jacqui Martin/Shutterstock
Page 152 (left): © Steven F. White
Page 152 (right): © MacArthur Foundation
Page 153: © weha/Shutterstock
Page 158 (top left): © Neptalí Ramírez Marcial/iNaturalist
Page 158 (bottom left): © Yarelis Benavidez
Page 158 (right): © superbphoto95/Shutterstock
Page 161: © Ben Kamm
Page 165: © guentermanaus/Shutterstock
Page 168: © Esthela Calderón
Page 171: collection of David Hornung; photo © Esthela Calderón
Page 175: © Yanti Lin/Shutterstock
Page 179: © Neil Logan
Page 182 (top and bottom): © Thibaud Aronson
Page 184-5: © Daniel Prudeck/Shutterstock
Page 191 (top and bottom): © Ben Kamm
Page 195: © guentermanaus/Shutterstock
Page 200 (top): © Ben Kamm
Page 200 (bottom): © Neil Logan
Page 201 (top): © Tom Baldwin
Page 201 (bottom left and right): © Ben Kamm
Page 205: © Gusmano Cesaretti
Page 205 (bottom): © Steven F. White
Page 207: © Esthela Calderón
Page 210 (top and bottom): © James Benefield
Page 217: courtesy of Yvonne Negrín, Wixárika Resource Center, Berkeley, CA
Page 219: © Nikitin Alexander/Shutterstock
Page 222: © R. K. Srimanee/Shutterstock
Page 227: image courtesy of Strictly Medicinal Seeds
Page 231 (top left): © Esthela Calderón
Page 231 (top right): © Camilo Jarquín Calderón
Page 231 (bottom): © Martin Pelanek/Shutterstock
Page 236: drawing courtesy of Ben Kamm
Page 236 (centre): © Ben Kamm
Page 238: © Curioso Photography/Shutterstock
Page 241 (top left and top right): © Ben Kamm
Page 241 (bottom) © Erik González
Page 246 (top): © Glenn H. Shepard, Jr.
Page 246 (bottom): © Timothy Paine
Page 247: © Riley Fortier
Page 249: © Thomas Laisné, Contour RA/Getty Images
Page 253: © MTN2705/Shutterstock
Page 255 (left): © Esthela Calderón
Page 255 (right): © C. A. Hill

We gratefully acknowledge the granting of permission to use these images. Every reasonable attempt has been made to identify and contact copyright holders. Any errors or omissions are inadvertent and will be corrected in subsequent editions.